Saving the Earth as a Career

Saving the Earth as a Career

Advice on Becoming a Conservation Professional

Malcolm L. Hunter, Jr.
University of Maine, Orono, Maine, USA

David B. Lindenmayer
Australian National University, Canberra, Australia

Aram J. K. Calhoun
University of Maine, Orono, Maine, USA

Second Edition

WILEY

Contents

Preface

This book took its embryonic form during the 2005 Society for Conservation Biology Conference at the University of Brasilia as we sat in a quiet spot waiting to see a swallow-tailed hummingbird and talking about joint projects for an upcoming sabbatical during which David would be hosting Aram and I at the Australia National University. We covered many possibilities, but the one that captured our sharpest interest was writing a book for young conservation professionals. No doubt we were inspired by the success of the Brasilia conference in attracting hundreds of young people from scores of countries around the globe. The key challenge would be to write an accessible, engaging book that would counsel and encourage people who are on the road to becoming conservation professionals, and not degenerate into a diatribe from some grumpy old people. We don't feel old, but we have been around long enough to have witnessed a fair amount of anxiety, stress, and inefficiency and we hope we can offer advice about how to avoid some of the pitfalls.

Being surrounded by people from all over the world in Brasilia led to our decision to write for a global audience despite the challenges of dealing with many different systems of education and employment. For example, key aspects of the educational experience at some universities, such as internships, comprehensive exams, and even course work, are totally unknown at other institutions. Even writing in one language has some difficulties because of the different versions of English. Notably, we have used the somewhat awkward term faculty member because professor, the simple synonym used in North America, is reserved for senior faculty members or department heads in many other countries. On the other hand, we have used the North American graduate student instead of the more widespread postgraduate student because it is shorter. Similarly, we chose advisor over supervisor because it is shorter and friendlier, and so on. We have tried to stay on the high ground of broad principles rather than wade into prescriptive specifics. This has allowed us to be both relatively cosmopolitan in our approach and to keep the book slim and inexpensive. Keeping the book to a manageable size was also behind our decision not to attempt a comprehensive synthesis of the relevant literature. Thus the ideas expressed here are predominantly our thoughts on an issue, with some modest tempering provided by our reviewers and some limited reading of the literature. We have provided some Further readings to provide a starting point for readers who want to dig deeper.

Although university students are our primary target audience the book will also be of some use to older readers (those contemplating a shift into conservation work from their current careers) and younger readers (pre-university students who are still planning their educational and career paths). The book takes a roughly chronological approach, beginning with first-year undergraduates, on through the years of graduate work, and into the job market. We have tried to encompass the entire span of conservation careers but inevitably our backgrounds in the biological sciences have tilted the book a bit in that direction rather than toward the social sciences and humanities.

We first want to thank all of the students with whom we have interacted over the years for we have learned a great deal from them and derived many hours of enjoyment from being with them. This book is dedicated to them and as a gesture of recompense we are donating all the royalties to student activities of the Society for Conservation Biology. We also want to thank all the people who reviewed drafts of the chapters and who provided many of the short anecdotes that appear here in italics: Darron Collins, Annika Felton, David Johns, Andrew Knight, Ewan Macdonald, Rebecca Montague-Drake, Mark Anderson, Marianne Asmussen, Jacolyn Bailey, Guy Baldassare, Andrew Balmford, Frederic Beaudry, Paul Beier, Sean Blomquist, Nora Bynum, Steve Campbell, Richard Cowling, Scott Delcourt, Megan Gahl, Yrjö Haila, Susan Jacobson, Rick Knight, Gary Meffe, Ed Minot, Trinto Mugangu, Fiona Nagle, Steve Norton, Pilar Palacios, David Patrick, Lindsay Seward, Javier Simonetti, Steve Trombulak, and David Wilcove. Although this is a personal account it has been very useful having reviewers tell us where they think our opinions fall outside the mainstream. We particularly thank the first six people mentioned for reading the entire volume and Bruce Doran for turning our suggestions into engaging cartoons.

Preface to the second edition

Almost a decade later, Aram and I returned to Australia National University for another sabbatical with David, a perfect impetus to revisit the first edition of this book. We found that what had seemed like sage advice in 2006 had stood the test of time, but inevitably some changes were still needed. Notably, the progress of technology – witness the explosion in social media – led to new material. More importantly, the ever-growing popularity of graduate degree programs that are focused on course work, as opposed to the more traditional emphasis on thesis research, led to extensive changes. As we head into production we need to pause to thank the many people who offered us useful feedback on the first edition, as well as those who provided new examples and other input: Britt Cline (who reviewed the entire book with a youthful eye), Kathleen Bell, Karen Bieluch, Yael Calhoun, Kristine Hoffmann, Meredith Kirk-Lawlor, Lachlan McBurney, Bridie McGreavy, and Thea O'Loughlin, Renee Mullen, David Owen, Scott Simon, and Randi Trask.

Malcolm L. Hunter, Jr

Read this road map before you begin

If you arrived here without reading the Preface, you are probably one of those people who hates to read instruction manuals, but take a few seconds to read this and you may save yourself hours of reading chapters that are not particularly relevant to you.

- If you are an undergraduate student who is considering further degrees en route to a career in conservation, then you can, in the words from *Alice's Adventures in Wonderland*, "Begin at the beginning and go on till you come to the end: then stop."

- Undergraduate students who will not pursue further degrees will still find Chapters 1, 2, 9, and 10 of use. If you are doing a senior or honor's thesis, or any major independent project, then Chapters 6, 7, and 8 will be helpful too.

- If you are in a master's degree program you can skip the first four chapters for now, perhaps returning to Chapters 3 and 4 if you contemplate continuing for a doctorate degree.

- Doctoral students can start at Chapter 5.

- If you are currently in another career but considering a shift into conservation work you may want to focus on Chapter 1, and skim the rest of the book to assess the route you are contemplating.

Recognizing that many readers will pick a different set of chapters to read there is a little bit of redundancy (e.g., writing introductions for a proposal and for a scientific paper are covered separately), but not enough to make reading inefficient.

1 Is this the right career for you?

Have you wanted to do conservation work since you were a small child, or did the allure of being a ballet dancer or firefighter attract your earliest passions, with conservation work coming to the fore much later in your life? What is it about conservation work that attracts you: perhaps being on the front line to conserve the natural world, perhaps being able to spend your days working outdoors in special places? In this chapter we explore what it means to be a conservation professional and then consider some of the basic preparations necessary to travel down this road.

WHAT IS A CONSERVATION PROFESSIONAL?

When you think of an archetypal conservation professional, what do you envision? Perhaps a park ranger, responsible for ensuring that nature is conserved in a large tract of wild country, or a wildlife manager working with local

CONSERVATION REQUIRES PEOPLE WITH DIVERSE TALENTS

communities to restore populations of an endangered species. Certainly these people are well represented under the umbrella of conservation, but they are not alone. For example, although people educated in the natural sciences traditionally dominated conservation work, the role of people with backgrounds in the social sciences and humanities is now very large and still growing. If we define conservationists as people who protect the natural world from misuse and who oversee the wise management of natural resources, then it is nearly impossible to distinguish them from environmentalists, who seek to maintain and improve the environment, and ecologists, who study the interactions among humans, other organisms, and their environments. Thus, conservation professionals would also include a policy specialist employed by an environmental advocacy group to influence legislation, an academic who studies how ecosystems function and shares that knowledge with students, a government official who monitors water and air quality, and many other people. The umbrella stretches from those who actively support careful use of natural resources (for example, most foresters, game managers, fisheries managers, and range managers) to preservationists who strive to protect nature from human intrusions. This assemblage of people may seem very diverse, but compared to the population at large they are likely to share broadly similar values and interests. This is not to say that they are like peas in a pod. Indeed, intra-family squabbling does occur, for example when one conservation professional sees a particular forest primarily as habitat for a rare reptile and secondarily as a source of timber, while another has reversed priorities. One of the roles that social scientists play is providing systematic assessments of how and why such values differ.

Also under the umbrella you will encounter people as diverse as a civil engineer who designs hydrological systems for wetland restoration, a business person whose company recycles solid waste, a lawyer who writes environmental legislation, or a medical researcher who specializes in human diseases that are caused by environmental degradation. Some people would argue that these people are not conservation professionals. Is it the nature of the specific work that is important (writing legislation compared with writing a park management plan) or is it the ultimate impact on nature and natural resources? What about someone who works in finance, fund-raising, or information technology for a conservation organization? Their day-to-day work may be entirely removed from the natural world, but they may make a larger contribution to the success of the organization than a field biologist working for the same group. Box 1.1 provides a more comprehensive view of the issue, but even it does not cover all the possibilities.[1.1]

We could argue in circles about how to draw a line around who is or is not a conservation professional, but the distinction is not really important. The good news is that there is enormous latitude for developing a career that meets your skills and aspirations *and* makes a major contribution to conservation. It is important that you begin with some serious introspection. Are you a person who thrives working with a group of people from diverse backgrounds to solve a complex social problem, or are you more content sitting alone at a computer, or traveling around remote terrain gathering data with one other colleague? These and many other predilections are easily accommodated in a conservation career.

This book will be most relevant to the conventional conservation professionals who dominate the center of the umbrella, but at some level it will be of use to everyone who aspires to work under the umbrella, or near it. Wherever you fit, you will get where you are going more readily with a strong education, often a graduate degree, and thus your formal education is a major focus of this book, but we also address your informal education through various experiences.

CONSERVATION CONTRIBUTORS

Many people care about conservation issues profoundly but are not in a position to become conservation professionals. Nevertheless, they can integrate conservation into their work in some fashion. For example, painters, musicians, and other artists are well known for bringing nature and conservation into their work as a source of inspiration and as a vehicle for expressing their values, often reaching a broad audience. Indeed, when you consider that a janitor who uses environmentally safe cleaning products and recycles refuse is making a noteworthy difference, it is clear that everyone can integrate conservation into their work to some degree. Simply donating some of the monies generated by work is one of the easiest and most important contributions.

Millions of people undertake conservation work on a volunteer basis by spending their weekends taking children on nature walks, surveying bird populations, collecting water-quality samples, planting native vegetation, and so on. These activities are so rewarding – providing both recreation and a sense of purpose – that they are often the first step in leading people into conservation work as a full-time career. Some of these people reach this decision long after their days of youth have passed, but not too late to make a career change. Every year many middle-aged people give up their current

Box 1.1 A classification of some employment opportunities for conservation professionals

The matrix below depicts some of the major kinds of jobs that conservation professionals hold and the major types of organizations that employ them. The Xs indicate types of position commonly held by many people, the asterisks indicate posts held by modest numbers, and blanks indicate positions that are rare or non-existent.

	Government agency	Environmental NGO*	Educational institution	Consulting firm	Natural resource industry	Zoo, garden, aquarium, museum	Self-employed
Biologist	X	X	X	X	X	X	+
Earth scientist†	X	+	X	X	X	+	+
Educator	X	X	X	+	+	X	+
Engineer	X		+	X	X		+
Information technology	X	X	X	X	X	+	+
Lawyer	X	X	+	X	X		+
Policy, planning, administration	X	X	X	X	X	X	+
Pollution technology	X	X	+	X	X		+
Social scientist‡	X	X	X	X	X	+	+

*NGO stands for non-governmental organization and in this context refers to private, not-for-profit environmental or conservation groups such as the World Wide Fund for Nature and The Nature Conservancy.
†Earth sciences include geology, hydrology, climatology, soil sciences, and others.
‡Social sciences include anthropology, economics, sociology, psychology, and others.

(continued)

Box 1.1 A classification of some employment opportunities for conservation professionals (continued)

This classification is very coarse. To demonstrate how much detail is hidden let's consider just the first cell: government-agency biologist. First, there are hundreds of different types of biologist. We can separate them on at least four axes: (1) researchers, practitioners, and those who undertake both research and management; (2) taxonomy (e.g., entomologists, ornithologists, or lichenologists); (3) ecosystem type (e.g., marine, arid, or freshwater); and (4) systems focus (e.g., genetics, ecology, or veterinary medicine).

Next there are vast numbers of government agencies. State or provincial and national agencies are probably the largest employers of conservation professionals, but these posts also exist at lower levels of government (e.g., municipalities and counties) and in international quasi-governmental organizations like the United Nations. Some of the government agencies that employ conservation professionals will have words in their titles that you would expect: Conservation, Natural Resources, Environment, Lands, Parks, Outdoor Recreation, Forest, Agriculture, Marine Resources, Fisheries, Wildlife, Energy, Mineral Resources, Soil, Air, and Water. Others might surprise you, such as Public Works, Defense, and Health.

Finally, there are many positions that are not covered by this matrix at all. For example, organizations as diverse as charitable foundations, ecotourism businesses, professional societies, and manufacturers of pollution-control or energy-conservation equipment employ some people who can be considered conservation professionals. Furthermore, people who are employed by conservation organizations in support roles – for example, accountants and human-resource specialists – may not be conservation professionals *per se*, but they certainly contribute to the cause of conservation and are often drawn to these posts because of their conservation values.

work to return to university and earn a degree that will allow them to chart a new path as a conservation professional. This is not an easy path to take, at least for people with major responsibilities like raising children or paying a mortgage, but it is possible.

> *We had a graduate student who earned her PhD in wildlife ecology following a 10-year career working as a doctor specializing in internal medicine. Her strong background in physiology allowed her to complete a successful research project on the physiology of bears during hibernation.*

People who have reached retirement age are less likely to pursue university degrees, but their financial independence often allows them to undertake volunteer conservation work on a nearly full-time basis.

> *My father retired at 54 and became president of the local bird club, a senior councillor in our local conservation non-governmental organization (NGO), and a major contributor to regional conservation policy development. He ended up working 50 hours per week after he had "retired" and has made a huge contribution over the last 15 years.*

In short, if you decide that a conservation career is not right for you, there are ways to still be a "conservation amateur" and make a solid contribution. Conversely, if trying your hand at being a conservation volunteer convinces you that you should become a conservation professional, then you can still make the switch late in life.

DIVERSE COMPENSATIONS

Need a house with an Olympic-size pool? Want to vacation for a month on the French Riviera? You have probably already figured out that most conservation professionals are not rolling in excess money; that their job satisfaction comes more from the rewards of the work itself than vast financial wealth. But let's explore this issue a bit more deeply. You may recall from an introductory economics course that the balance between supply and demand determines prices and this implies that wages would be low if the demand from conservation employers were easily met with the supply of people willing to take a conservation job. However, despite the many attractions of conservation work, the pool of conservation professionals is not completely flooded because of the specialized education and skills usually required, typically at least 4 years at university, and often 2 or more years of postgraduate education, coupled with some demonstration of professional experience. Thus, the good news here is that for those who do attain the necessary education, salaries are definitely respectable.[1,2] You may not get rich, in part because most conservation jobs are with government agencies and non-profit organizations rather than in private business, but you should be able to have a perfectly fine standard of living while enjoying a strong sense of personal fulfillment.

If you want to be a conservation professional and be *more* than financially comfortable, there are routes to pursue. In particular, conservation jobs in the private sector, notably with consulting firms, often pay very well, and leading a conservation group almost always pays better because of the added responsibility of managing large budgets and many employees.

Of course, less tangible benefits, notably job satisfaction, are often paramount for conservation professionals. Simply put, it is extremely rewarding to feel that you are working to make the world a better place. However, do not get the idea that every day you will feel triumphant. In the world of conservation, David often loses to the Goliath of insistent forces that degrade our planet, but you can always go to sleep at night knowing that you are striving to make a difference. And that is priceless for people who are passionate about conserving the natural world. If you lack that passion then you may get along as a conservation professional, but you will probably not flourish.

In the world of conservation, David often loses to the Goliath of insistent forces that degrade our planet, but you can always go to sleep at night knowing that you are striving to make a difference.

One of the key benefits of being a conservation professional is that the work is sometimes truly fun, pure and simple. As you know, not every conservation worker spends every day traveling to beautiful places to interact with creatures that other people seldom see, but such days are a part of conservation work and they go a long way to compensate for the time spent in front of a computer or attending meetings that are a large part of most jobs, including conservation.

> *Some of the most memorable moments of my life have come through my employment: for example, while diving in the Caribbean to survey corals or searching for red pandas on the slopes of the Himalayas. Being a conservation professional has also opened doors to be with colleagues during their work; for example, joining them to watch a dozen male right whales churn the waters of the Bay of Fundy, or a gorilla family foraging on bamboo in the Virunga Mountains.*

Of course, these experiences are the icing on the cake, and aspiring conservation professionals must realize that they are greatly outnumbered by hours hammering away at a keyboard for most of us. Still, compared to many jobs that have a large proportion of drudgery, conservation work stands out as quite attractive.

LOCATION, LOCATION, LOCATION

Do you like to pick up a cup of gourmet coffee on your way to work each morning? Or do you prefer to live 50 kilometers beyond the nearest electrical outlet, let alone espresso machine? Both lifestyles are entirely compatible with being a conservation professional because this career provides opportunities across the entire spectrum of urban and rural, even wilderness, settings. This is not true of all professions: if you are a piccolo player you will need dozens of other musicians to form an orchestra, or if you want to be a rancher you will need lots of open space for your herd. Compared to the population as a whole, conservation professionals are more likely to be in remote places than your average citizen. However, looking just at the distribution of conservation professionals, more of us work in cities than in the countryside because most government agencies and conservation groups have their offices in cities. Fortunately for these folks, their work may take them to places where they can enjoy wild nature for a limited period and then they can return to the day-to-day comforts of gourmet coffee.

THE GRASS IS ALWAYS GREENER.....

If you have your heart set on living in a particular rural spot, perhaps the island or valley where you grew up, then a conservation career will probably be more constraining than a common occupation that is in wide demand, such as teaching or nursing. Perhaps you can create your own conservation job – setting up eco-tours of your valley for example – but it will take considerable initiative. On the other hand, if you are a native of Madrid, Montreal,

Miami, or Melbourne and want to stay close to your roots, most cities have jobs for a fair number of conservation professionals. The bottom line is that even if you are choosey about the type of environment you wish to live in, a conservation career can easily fit your needs; if you have a specific rural place in mind you may need to be creative or lucky. Fortunately, new communication technologies (e.g., video conferencing) are creating ever more flexibility about where people work.

YOUR IMAGE

The most important person that you need to satisfy is yourself. If you are uncomfortable with your career choice – in particular, if you do not respect yourself and your work – then you are on the wrong path. This is not an issue for most conservation professionals; they are usually very proud of their work and justly so. But what do you do when someone challenges your choice of a career? Many conservation professionals, particularly those who work in beautiful places, regularly hear questions like, "They pay you to do this?" or "When are you going to get a real job?" Sometimes the motivation behind the question is envy. Many people are bored or disillusioned with their current jobs and would love to be able to switch to an exciting, worthwhile career. For these people, the best response is to give them an honest assessment of conservation work, which has its rewards but is not all milk and honey.

Sometimes these questions reveal a lack of respect for conservation work. People who do not value the natural world, people who measure their career success in terms of their salaries, may not understand the motivations of a conservation professional. They are likely to be thinking, "Why would someone compromise their earning power in a career that may stand in the way of commerce and other human enterprises?" You are not likely to change the values of such a person in a brief conversation, but there is no harm in trying. You might garner a modicum of respect by making an analogy to the widely respected medical profession, and pointing out that conservation professionals are doctors for the Earth.

Conservation professionals are doctors for the Earth.

TALK AND EXPERIENCE

Reading this book is analogous to having a long, in-depth conversation with three people. Collectively we have quite a bit of experience, but nevertheless the views expressed here are the opinions of only three people, somewhat refined by those whom we asked to review drafts of the book. It is a good idea to solicit

more opinions about the topics broached in this chapter and in all the chapters that follow. Talk to as many conservation professionals as you can about what they like and dislike about their careers and how they got started on their career paths. Ask your teachers about other students with similar interests whom they have counseled in the past. Talk to other students who are also contemplating a career in conservation, especially graduate students who are further along the road than you are. If you hear different opinions, try to sort out what lies behind these differences, perhaps returning to someone you talked with earlier to ask for clarification.

Talk to as many conservation professionals as you can about what they like and dislike about their careers and how they got started on their career paths.

The folk wisdom captured in phrases such as "talk is cheap" and "you need to walk the walk, not just talk the talk" suggests that you should go beyond chatting about conservation as a career over a cup of coffee. Ideally you would spend some real work time with a few conservation professionals. In the next chapter we will discuss the importance of finding substantial summer jobs, but there are some baby steps that you can take in that direction, simply to give you a better assessment of whether this is the right career path. For example, you might be able to tag along with a conservation manager for a day or two doing field work. Perhaps you can volunteer to spend a couple afternoons a week helping out in the office of an environmental group. You might end up stuffing envelopes, but just being in that work environment will give you subtle cues as to what it is like to be a regular employee there and the chance to develop relationships with potential employers.

Whatever approach you take, sorting out your career options is an important undertaking. It certainly merits taking the time to gather information, opinions, and ideally experiences from a diverse array of sources.

Further readings and notes

1.1 A number of books provide a broad overview of the kinds of careers that fit under the umbrella of conservation or the environment. Before making a purchase read reviews as some of them are narrower than you might surmise from their titles, and most are not global in scope.

Cassio, J. and A. Rush. 2009. *Green Careers: Choosing Work for a Sustainable Future*, New Society Publishers, Gabriola Island, B.C., Canada.

Deitche, S.M. 2010. *Green Collar Jobs: Environmental Careers for the 21st Century*, ABC-CLIO, Santa Barbara, California, C.A.

Environmental Careers Organization. 1999. *The Complete Guide to Environmental Careers in the 21st Century*, Island Press, Washington D.C.

Environmental Careers Organization. 2014. *The ECO Guide to Careers that Make a Difference*, Island Press, Washington D.C.

International Labour Organization. 2011. *Skills for Green Jobs: A Global View*, United Nations, Geneva.

Also note that New Society Publishers has two books with a more entrepreneurial approach to careers in conservation: *Making a Living While Making a Difference: Conscious Careers in an Era of Interdependence* (2007) and *Ecopreneuring: Putting Purpose and the Planet before Profits* (2008).

Virtually all organizations that are active in conservation have websites that you can peruse to get some flavor of the work they do, and most environmental non-governmental, non-profit organizations (NGOs) also have magazines and newsletters.

1.2 *The Complete Guide to Environmental Careers in the 21st Century*, see note 1.1, gives information on salaries for the USA but you will have to extrapolate from 1999.

2 Establishing an undergraduate foundation

Deciding that you want to devote your career to conservation is a good start up the mountain, but of course it is not sufficient in itself. You have to be willing *and* able; a good do-er as well as a do-gooder. You need skills and knowledge as well as commitment and purpose. In this chapter we will focus on some ways to start developing those attributes while you are an undergraduate student, or an older person seriously considering a switch to conservation. If you are successful, you may well be adequately prepared to start a conservation career with a bachelor's degree and a modicum of extracurricular or professional experience, but as we will cover in Chapter 3, we generally recommend using these preparations as a foundation for advanced education as a graduate student. Thus many of the topics covered below are examined more thoroughly in later chapters from the perspective of a graduate student.

UNIVERSITIES AND DEGREES

There is a dizzying array of institutions where you can pursue higher education in conservation as broadly defined in this book.[2.1] Indeed, tertiary institutions that do *not* have some form of conservation or environmental program are quite uncommon, especially when you consider the vast array of disciplines that contribute to conservation. There are even some places (e.g., the Norwegian University of Life Sciences, the College of the Atlantic, and the College of Environmental Science and Forestry of the State University of New York) where this discipline is a dominant theme for the entire institution.

Saving the Earth as a Career: Advice on Becoming a Conservation Professional, Second Edition. Malcolm L. Hunter, Jr., David B. Lindenmayer, and Aram J. K. Calhoun.
© 2016 John Wiley & Sons, Ltd. Published 2016 by John Wiley & Sons, Ltd.

The list of possible degrees you could pursue is also seemingly endless, with many larger universities offering dozens of choices that could be appropriate for an aspiring conservation professional. Some will have obvious titles, such as Natural Resources Management, Ecology, and Environmental Studies, whereas others might surprise you, such as Civil Engineering, Law, or Resource Economics. Most will be bachelor's degrees but there are some associate degrees and diplomas that also can lead to professional employment in conservation.

We have assumed that most readers of this book have made their first step towards a career in conservation by having enrolled in an undergraduate degree program that is somewhat related to conservation, more likely public policy or biology than music or mathematics. However, if you are still in the market for an undergraduate program, you will find some advice, useful in broad brush strokes, on selecting universities and degrees in Chapter 3, where we cover the topic from the perspective of an undergraduate selecting a graduate program. Later in this chapter we address the issues faced by a person who is considering switching to conservation from a different career track, either during a university degree or long after earning one.

COURSE WORK

With conservation covering such a breadth of topics it is not possible to offer a long, detailed list of courses that every aspiring conservation professional should take. Nevertheless we feel on reasonably solid ground suggesting that three subjects should lie at the foundation of almost every conservation curriculum, even for people who end up far from the scientific side of conservation (e.g., as an environmental lawyer). First, there is biology; clearly you need to understand life if you want to conserve life on Earth. Second, ecology is essentially the environmental side of biology and you need particular strength in this topic. Third, you will want to take one or more courses on conservation issues and how they are solved, which will likely involve the intersection of ecology with the human dimensions surrounding natural resource management.

The conservationists who have the biggest impact on the world are not hermits. They are out in public communicating their knowledge and perspectives. Consequently you should take courses that will hone your skills in both written and oral communication. These may be courses in speech, writing, photography,

The conservationists who have the biggest impact on the world are not hermits. They are out in public communicating their knowledge and perspectives.

journalism, or social media, or courses within your core curriculum that require writing multiple papers, generating social media content, and speaking in front of the class on a regular basis. If you are much more comfortable with one mode of communication than the other, work on your weakness. Although universities often stress academic writing, being effective with other types of written, visual, and oral expression and a variety of outlets may allow you to reach a broader audience.

Balancing the need for both depth and breadth is often an interesting challenge. For example, you may want to take many ecology courses – such as behavioral ecology, marine ecology, forest ecology, and physiological ecology – but at some point this will constrain your ability to pursue breadth, most notably some social science courses – such as resource economics – that will make you more effective at dealing with human institutions. Conversely, if you are focusing on the social-science side of conservation, you should consider taking a course like ecosystem dynamics or ornithology rather than another economics class.

To some degree, these decisions will be made for you because graduating with a particular degree requires working to fulfill certain university requirements, although the latitude allowed varies enormously among universities. Whenever you have choices to make it is obviously wise to look beyond the immediate issues (e.g., is your best friend taking the course?) to how well a course will serve you down the road. For example, a macroeconomics class may be a tough slog now, but if later you apply for a graduate assistantship to study resource economics it will serve you well. Some conservation students shy away from classes in math, economics, and statistics, but these provide an important background for many careers in conservation.

Your advisor (known as a supervisor in some parts of the world) will know the options for your university and can help you find your way through all these decisions, so be sure to express your interest in becoming a conservation professional. Getting good advice may entail switching advisors; for example, if you are in a biology department where most of the advisors are focused on preparing students for careers in medicine, or in a civil-engineering program where building roads and bridges completely overshadows construction of water-treatment systems. Having a conservation-oriented advisor is especially useful for looking at issues that go beyond the university, such as meeting the professional certification standards that some professional societies have developed, establishing a network of future colleagues, or pointing you toward an opportunity for practical experience.

COURSE PERFORMANCE

You may feel that you have spent most of your life striving to earn good grades but, as you know, this is not just an exercise in pleasing your parents. The people who select employees and graduate students believe that earning good grades requires some combination of intelligence and hard work and therefore grades are a significant consideration. With poor or even mediocre grades your application may not make it past the first filter. Fortunately, it is fairly common for

ATTITUDE PLUS APTITUDE EQUAL SUCCESS

students to have a lack-luster overall average grade but one that improved steadily through their undergraduate career, or one that is much better for courses in their discipline. If this is the case for you it is worth highlighting this pattern in any application letters.

Course performance may be measured primarily by grades, but not exclusively. Every faculty member knows students who are not at the top of the class

Your social skills are very important to your professional advancement and this begins in the classroom, where your instructors and fellow students are essentially your first colleagues.

in terms of grades, but who outshine everyone else in terms of their attitude and work ethic. They are always on time for class, ask good questions, undertake assignments with enthusiasm, greet people warmly in the halls, volunteer for extracurricular activities, and so on. When it comes to references for a job, people with these kinds of characteristics may be just as likely to be recommended as someone who scored highest on the last exam. Of course, your best bet is to earn superb grades *and* be a pleasant, enthusiastic person. The point is that your social skills are very important to your professional advancement and this begins in the classroom, where your instructors and fellow students are essentially your first colleagues. Although excellent grades and classroom social skills are important, they are not sufficient to carry an application that is very weak in other respects, as we will see in the next section.

EXPERIENCES OUTSIDE THE CLASSROOM

Short-term jobs

Not so long ago most people learned their key skills informally – how to grow a garden, cook a meal, or repair an engine – either from their family or as an apprentice to a professional. Acquiring skills outside the classroom is also a critical part of becoming a conservation professional. For example, here is a list of some general skills that you may be more likely to learn in a short-term job than in class: working both independently and as part of a team, problem-solving, communicating with the public, recording data meticulously,

functioning well under adverse conditions (notably unfriendly people, bad weather, and biting insects), reading a map, and observing and identifying biota. Specialized skills are also more likely to be learned and refined during weeks of employment than in an afternoon lab session: managing a database, using geographic information systems (GIS) and global positioning systems (GPS), interviewing people, extracting DNA samples, and radiotracking animals, to name just a few of many.

Good jobs are the primary route to robust experiences and an impressive résumé. For most undergraduates these will come during the long breaks between terms because in 2–4 months you can really immerse yourself in a job and travel far to give yourself some geographic diversity.[2.2] Working weekends and after classes is another option that can diversify your work portfolio. Part-time work during the school year is particularly useful for breaking into the summer job market as a first-year student. If competition for jobs is tight you may need to start with a volunteer position, such as helping with a professor's research or working without pay one afternoon per week for a local environmental group or government agency.

If an unusually attractive opportunity comes along – for example, a chance to spend the winter term working on an environmental project in another country – you should even consider suspending your regular course work for a term or two. Very few people will question the wisdom of taking an extra year to earn your undergraduate degree if you can show how that extra year clearly advanced your development as a professional.

It is also common for undergraduates to seek short-term jobs after graduation, before going on to a major commitment. A year or two of working on a number of jobs in the conservation field, or one longer-term job with diverse responsibilities, will make you much more competitive for a graduate position or long-term job. It may also give you a much-needed break from course work before starting graduate school. Consider working far away from home; it will probably be easier to pursue adventurous opportunities at this stage in your life than at any other time.

You may have the opportunity to repeat the same job at a later time. This is obviously an advantage to your employer, who will get a proven, seasoned worker, but it is probably less than ideal for you unless it puts you on track for the perfect long-term job. We would generally recommend it only if you will be given major new responsibilities (e.g., being a team leader) and you can diversify your overall set of job experiences in some other manner (e.g., a long-term weekend job or working after graduation).

In summary, building your repertoire of skills through diverse, relevant work experiences in the conservation field is critical to your success. Students who focus all their time on course work and then spend their summers seeking the highest-paying job to pay off car loans and tuition bills are being short-sighted. You must have some work experience, preferably many months' worth, but at least one long period and numerous weekends.

Building your repertoire of skills through diverse, relevant work experiences in the conservation field is critical to your success.

Integrating course work and jobs

Because skills gained on the job are so valuable, many universities formalize this process by offering students the opportunity to receive university credit for job experiences, thus allowing you to take time of out of the academic year without delaying graduation. These go by many names, such as internships, service learning, or field experiences. As we have emphasized, these experiences are extremely valuable, although one could debate how much more value is added by formalizing them as a university course, at least if this only involves keeping a professional diary. The key feature is that they may open a special opportunity that your university has negotiated with an employer, initially for a short-term post and then possibly for long-term employment. For example, some government agencies sponsor "pathways" programs from universities to long-term employment.

Sometimes the integration of coursework and practical experience comes in the form of an independent project: your chance to design, execute, and present your own work under the tutelage of an advisor. This is a fantastic opportunity for out-of-classroom enrichment that is highly analogous, on a smaller scale, to graduate-student thesis research. It is sometimes called an honor's thesis or senior thesis; in many countries students who aspire to graduate school will undertake an optional final year devoted to an honor's project. A successful project, notably one that generates a publication in a professional journal, will definitely help pave your way into a first-class graduate program. Even if a thesis-based graduate program is not your ultimate aim, an independent project may still provide an invaluable opportunity to design and undertake your own work. If you do develop an independent project, you will find useful guidance in Chapters 6–8.

Most undergraduate projects have to be squeezed in after regular classes and on weekends, but you may be lucky enough to find a summer job with latitude for you to undertake an ancillary independent project. For example, imagine that you were working as a research assistant on a project that involved checking pitfall traps for reptiles, and you proposed to your supervisor that you work on an ancillary project that would advance the goals of the overall research program: collecting and identifying the beetles that are also caught in the traps. Your supervisor would probably be more than happy to have you use the project's resources and offer you guidance in your special project.

Extracurricular activities

Very important social skills, such as learning how to lead a group, can come from membership in organized student activities. Of particular relevance to conservation students are university-based environmental advocacy groups and the student organizations affiliated with professional societies, such as chapters of the Society for Conservation Biology or the Soil and Water Conservation Society. Joining a professional organization and assuming a position of leadership is one of the very best ways to make a crucial shift in perspective: you need to stop thinking of yourself as a student studying a conservation discipline and starting thinking of yourself as a conservation professional who happens to be at the student stage of your career (see Chapter 7 for more on professional societies).

Stop thinking of yourself as a student studying a conservation discipline and starting thinking of yourself as a conservation professional who happens to be at the student stage of your career.

Another important way to develop yourself professionally is to spend time with people who have more experience, people who can serve as mentors and role models. Student organizations can be a great vehicle for facilitating this. For example, you could invite conservation professionals to make presentations at your meetings and invite them out for a meal.

One of the most useful activities of our student chapter of the Wildlife Society has been a program that links up local professionals with students who would like a mentor with whom they can discuss career options and other issues. It has proven very useful for the students and the professionals have thoroughly enjoyed the experience too.

Graduate students also can be excellent mentors. Being just a few years ahead of you, your questions and concerns are likely to resonate very clearly with them. One of the best ways to get to know graduate students is to volunteer to assist with their projects on weekends or after classes. Such work will often put you at the head of the line when paid opportunities arise.

If your department has a seminar series with invited speakers, try to organize your schedule so that you can attend these. Commonly graduate students and faculty members dominate the audience at these events, but you will almost certainly be welcome. Just being in the right place at the right time can lead to opportunities. If you have been attending seminars with a group of people every Friday noon for the whole term you may well be the first person who hears about the new opening for a summer assistant. In short, make yourself visible: talk to people in the conservation field.

Recreation

Even your recreation time can be a vehicle for becoming a more effective conservation professional. Are you interested in learning the techniques of scuba-diving, orienteering, or small-boat handling? How about various naturalist skills such as birding and botanizing, or communication skills such as recreational writing or setting up a website? Specialized skills that are fun to learn can open doors. Even just being able to stay comfortable and safe in remote situations and adverse conditions is an important skill that comes from spending lots of time in wild places.

One could even make a case that the most passive and ubiquitous of recreational activities – surfing the web – could further your development as a conservation professional if strongly oriented toward watching nature documentaries and keeping abreast of current events with relevance to conservation. However, most conservation professionals are very dismayed by the extent to which the public at large gravitates towards computers, televisions, tablets, and other electronic distractions. We strongly advocate getting out to experience the natural world directly, particularly if that involves honing your observation skills. There is a second issue hidden here: physical fitness. While most conservation work does not require strength and endurance beyond the fitness of the average individual, even desk-bound conservation professionals will be more effective if they are in good health and this goes hand in hand with an active, outdoor life.

Summary

In summary, your quest to become a conservation professional does not end at the classroom door. It should become thoroughly integrated into your life, stretching in a seamless fabric from academic work to weekend recreation to summer employment. This is not to say that conservation has to occupy 100% of your life – it is fine to enjoy a volleyball game or a visit to an art museum, for example – but you are not likely to be successful if you confine your career development to course work. In Chapter 5 you will find additional ideas on how to integrate your personal and professional life.

Your quest to become a conservation professional does not end at the classroom door.

REFERENCES

Historically, a person setting off for new places or new endeavors would solicit letters of reference from influential people. For example, Darwin's invitation to sail on the Beagle was orchestrated through letters among a network of his connections. Today, success in life is much more about *what* you know, than *who* you know, but letters of reference remain fairly important for one good reason: it is impossible to characterize who you are in a résumé and course transcript. Are your good grades the result of a Solomonic intellect or a Herculean capacity for hard work? Are you better working independently or as a member of a team? Are you a highly creative person who can develop a new line of thinking or a meticulous person who will carefully execute an existing protocol?

Usually you are asked to give the names of three people who are willing to be a reference for you, but it is probably a good idea to have at least four arrows in your quiver in case someone is unavailable at a crucial time. Be sure to include at least one faculty member who knows your academic capabilities and one employer who has seen you at work in a professional setting. Cultivate these people over an extended period so that they know you as a pleasant, multi-faceted individual, not just someone who gets good grades on exams. They will become part of your life-long social network, a key element in finding opportunities and capitalizing upon them.

When the time comes to ask people to serve as your reference, be prepared to give them a copy of your résumé, your letter of application, perhaps some key points you would like them to address, and any other items they need to have (e.g., reference form) or may wish to see (e.g., a copy of your transcript). Be sure to give them plenty of time in advance of the due date for any particular reference; at least 2 weeks, but preferably a month. If a person has agreed to be a reference for an extended period, offer to provide them with updates as you apply for specific positions so that they are not blind-sided when a reference request arrives.

You should also be aware that the conservation community is small enough that an informal system of checking on qualifications often operates. You may list Drs Black, White, and Gray as your references but if the employer knows Dr Green from your department, Dr Green may also get a phone call asking about you. In other words, it is wise for you to nurture good relationships widely and not just focus on people who are serving as your references.

STANDARDIZED TESTS

Standardized tests like the Graduate Record Exams (GRE) and Test of English as a Foreign Language (TOEFL) may seem like a dark step toward an Orwellian world where you are defined by your genetic profile and a few numbers, but they are an unavoidable part of pursuing higher education at many universities, especially in North America. Although many people are skeptical about how meaningful these exams really are, poor scores can sink a university application in the early stages, especially if you are competing in a large pool. Conversely, very good scores can sometimes open doors to generous stipends or scholarships. The significance of good test scores rises if you are a graduate of an institution that is not familiar to the people reviewing your application, and this is especially true if you are applying from overseas. You may have graduated in the top 10% of your class at the leading university in Paraguay or Ghana, but unless the standards of your university are well known to the reviewers this may not mean much. Within North America, hundreds of small colleges and universities are not widely known and so this issue can also be important when sending applications across state or provincial borders. The take-home message is this: prepare yourself for any required standardized tests. There are many resources, easily located on the Internet, for assisting you in this.

CHANGING COURSE

If you are reading this book while still wearing the hat of a student of music or nuclear physics you can certainly alter your compass, and it is better to do so sooner rather than later if you are convinced that a career in conservation is your goal. This change may be very easy. For example, in a university where students have significant latitude in their curriculum you could probably still switch from a general biology degree to a focus on conservation biology 2 or 3 years into your degree program and still graduate on schedule. On the other hand, going from physics to conservation might mean starting over from the beginning, especially at a university where standards are strict and there is no overlap between the conservation curriculum and the

Many people seek education that will prepare them to become a conservation professional after going down a different track for a while.

physics curriculum. You can start assessing what steps would be required to make a shift by comparing the courses you have taken with the requirements of a conservation degree. If there is a glimmer of hope, talk to a couple of faculty members in a conservation program; they may know some shortcuts that are not obvious to you. For example, at some flexible universities you might be able to create your own interdisciplinary program that bridged your original program and your new interest in conservation. Ultimately, as described above, you will need to do more than change your curriculum. Pursuing relevant jobs and extracurricular activities will need to go forward at an accelerated pace.

SWITCHING CAREERS

If you are a person who has already earned an undergraduate degree in another discipline, your preparations will diverge to some extent from those outlined throughout this chapter. In particular, rather than starting over with another undergraduate degree, you might have the option of going directly for a graduate degree in conservation after completing a much more limited set of courses, something that you could complete in a few terms. Obviously the details of this transition will vary enormously. For example, it would be much easier for someone going from a microbiology degree into conservation biology than for someone switching from art to environmental economics. In the latter case, you may well find it more prudent to pursue a second undergraduate degree, and then assess the desirability of graduate work. No matter what your background, it is likely to require a unique process to craft a fit between where you are and where you want to go. The best advice we can offer is that you try to find a conservation-oriented faculty member at a local university who would spend half an hour with you reviewing your background and goals and making recommendations. Ideally try to talk to two or three people so that you can evaluate a diversity of opinions.

Once you have charted a course and started attending classes, your circumstances will continue to be a bit different from the typical undergraduate. Of course it will still be important to earn excellent grades and standardized test scores to demonstrate that you have not lost your knack for academic work. Fortunately, many returning students perform much better on their second time around at university.

With respect to non-academic preparations, your work experience will probably have provided you with many generic skills, such as working with a team, being resourceful, managing a budget, or problem-solving. If you wish to pursue the field side of conservation work it may be more important to demonstrate that you have the stamina and tolerance for field work under difficult conditions and that your love for conservation work is not just a whimsical notion. In this case, you should consider seeking weekend employment in conservation, perhaps as a volunteer, or, better yet, take a leave from work to spend several weeks doing some approximation of the work you aspire to. Of course, these experiences may or may not convince you that switching careers is a wise idea.

There is a good chance that your recreational interests are already aligned with enjoying the natural world; if not, there is no time like the present to start. You may not feel comfortable becoming involved with student organizations if they are dominated by people 20 years your junior, but check them out first as you might be surprised. If student groups are not a good match, check out the community for local environmental groups, nature clubs, and other organizations.

I entered the environmental field after retiring from a career in the arts. Returning to school was a hard adjustment mentally and financially, but I feel like I'm going to be doing really good things in the world. By volunteering for committee work with a professional society I have developed a network that includes many leaders in my field, people that I can call upon when it is time to job hunt.

NEXT STEPS

Once you are feeling both willing and able, you are ready to pursue a long-term job in conservation or graduate-level education. For most students, we recommend graduate education as an important stepping stone to a career that will provide you with a rich set of opportunities, challenges, and responsibilities; we describe how to get started in Chapter 3. For some students an undergraduate degree will suffice and you can turn to Chapters 9 and 10, perhaps after skimming the intervening chapters to further evaluate whether you want to try your hand at graduate work.

Further readings and notes

2.1 We know of no central clearing house for universities that offer conservation degrees, but one of the largest databases is maintained by the Society for Conservation Biology conbio.org/professional-development/academic-programs. You can also search the Internet using "university" or "college" as a keyword along with various descriptors for your topical and geographic interests.

2.2 To find summer jobs, see Chapter 9 and the resources listed for it. For students and recent graduates in the US, the Student Conservation Association (www.sca-inc.org) is a specialized gateway for summer and temporary jobs. Operation Wallacea (opwall.com) provides students with conservation research experiences in over a dozen countries.

3 Selecting an educational program

What shall I have for dinner? What shall I do for a career? Life is full of decisions, some quite simple, some very complex, and obviously pursuing graduate education will take you to the complex end of the spectrum with four major elements – university, topic, advisor, degree – that often must be considered simultaneously. This is a challenge for anyone, completely overshadowing whether you have pizza or baked beans tonight. In this chapter we will review these four elements individually, and then see how they come together during the process of deciding where to apply.

THE KEY ELEMENTS: UNIVERSITY, TOPIC, DEGREE, AND ADVISOR

University

Many students spend their teenage years striving to earn excellent grades that will pave their way into a university that is widely regarded as prestigious. For selecting a graduate program, other issues beyond the perception of prestige become more important, especially the selection of an advisor, as we will describe below. Furthermore, from *your* perspective a university's reputation for producing conservation professionals is more significant than its overall stature. Evaluating this specific reputation is largely a subjective,

> *From your perspective a university's reputation for producing conservation professionals is more significant than its overall stature.*

Saving the Earth as a Career: Advice on Becoming a Conservation Professional, Second Edition.
Malcolm L. Hunter, Jr., David B. Lindenmayer, and Aram J. K. Calhoun.
© 2016 John Wiley & Sons, Ltd. Published 2016 by John Wiley & Sons, Ltd.

word-of-mouth exercise, but talking to faculty members, fellow students, and other conservation professionals about the relative merits of different institutions will be a helpful first step.

A university's website can also be mined for information.[3.1] Does the university offer an appropriate degree? Does the number of faculty and graduate students suggest that the program is neither too small nor too large? It is desirable to have many faculty working in your general area of interest not, for example, a single conservation person in a faculty dominated by molecular biologists. On the other hand, one can get lost in a huge program where people never even learn the names of many colleagues.

> *I was visiting a university to give a guest seminar and went to lunch with about 10 graduate students. I was shocked to discover the students on the left side of the table had never met the students on the right side of the table, especially given that those on the left were studying predation on songbirds and those on the left were studying predatory mammals.*

Conservation degree programs are often interdisciplinary in nature and this is a good thing, but it can exacerbate the anonymity issue if relevant faculty members are dispersed over multiple departments with their offices scattered across campus. The many faces of conservation also means that at most large universities a number of independent departments or programs will fall under the umbrella of conservation, and you need to evaluate all of those that are relevant to your interests more or less independently. This is necessary because one can have, for example, a first-class natural resources program and a mediocre botany program at the same university. The main point here is that it pays to probe around the internal framework for a given university, rather than basing your assessment solely on the overall prestige from university-ranking reports.

Geography can also play a role in selecting the ideal place for your education. If your academic interests are geographically broad, it would be wise to spread your experiences across the nation or even around the globe. On the other hand, if you want to remain geographically narrow, then it may be better to pursue all your education in the same region, although not necessarily at the same university lest you experience academic inbreeding. Some conservation students diversify their experiences geographically by undertaking project work far from the university. In particular many students attend an overseas or distant university to broaden their academic horizons, but then return home to undertake their project. This allows them to work on a conservation issue that is close to their heart and perhaps fosters or maintains relationships that will lead to a job with a local conservation organization. Seeking geographic diversity is probably wise, but the most important issue is whether or not a university is home to an excellent advisor who shares your topical interests, and determining this will take some digging beyond the website.

Topical area

Some people have a specific interest that is driving them to seek advanced education — for example, climate change mitigation or control of invasive plant species — whereas others have much broader interests, such as environmental economics or marine conservation. In any case, no university covers the entire breadth of conservation so your area of interest will constrain the range of places that are suitable for you. If your interests are broad, the list of universities will be long and they will be relatively easy to find. Just search the Internet using the term "university" and a couple words describing your topical area; try multiple versions such as "marine ecology," "marine conservation," and "marine resources."

If your interests are fairly specialized, you will need a different approach. You may be lucky and find a specific opening for a graduate student to study urban water policy, or whatever subject tickles your imagination, advertised on a website.[3.2] Of course, these advertisements will generate many inquiries so you should also search for less-conspicuous opportunities. For example, do not underestimate the utility of word-of-mouth networking with conservation professionals, professors, and graduate students. This process will also require a review of the academic literature using a database index such as Google Scholar or Web of Knowledge. In a few minutes you can have a long list of papers on urban water policy or any other topic. Combing these for potential universities will take longer. If you already have a short list of universities, you can screen for papers from these universities using the Address field of Web of Knowledge.[3.3] In this process, pay close attention to names of people who keep recurring, for this is where you are likely to first spot the name of some potential advisors.

Degree

BA, BS, BSc, JD, MA, MPA, MS, MSc, MPhil, PhD, DVM, DSc ... one step in pursuing your education is to decide where you are headed on the alphabet tree of formal qualifications in academia. There are conservation professionals working with just about every degree you can imagine and even a few with only minimal formal education (perhaps recruited from a village to be a liaison with the local community). That said, for many of the people who are likely to be reading this book, we would recommend some type of graduate degree because it will open the door to a large portion of the conservation jobs that will provide a rewarding career. Some people will be happiest with the kinds of jobs that are available with a bachelor's degree and some professional experience. Indeed, not too long ago a bachelor's degree was the norm, but in many organizations a sort of academic inflation now constrains opportunities for advancement if you only have a bachelor's degree.

Choosing between graduate degree levels – usually a master's degree versus a doctorate degree – is often not easy and it is hard to offer generic advice because the relative availability of these degrees varies among countries. In some countries master's degrees are the norm (e.g., the USA and Canada), but elsewhere they are often overshadowed by doctoral programs (e.g., the

UK and Australia). It is safe to say that a doctorate will usually be necessary for those who want a career in research (except as someone else's research assistant) or university-level teaching. In countries where master's degrees are readily available, people often decide to earn a master's degree first and then consider whether or not to continue for a doctorate. This is probably a prudent approach for most people, but if you are 95% certain that you want a career in research or academia then going directly for a doctorate may let you reach your goal a year or two sooner. At many universities it is possible to switch between master's and doctoral degrees in midstream, but some discourage this or may allow it in only one direction, so be sure to ask about this option if you are wavering between a master's or doctoral degree.

There are two basic types of master's degree, generally best distinguished by whether or not they require a thesis. Thesis degrees (often called MS or MSc for Master of Science or MA for Master of Art) put more emphasis on undertaking an independent research project than on course work. Degrees without a thesis are often called "professional degrees" because they emphasize the course work you need to go directly into a particular profession. Professional degrees may require a special project or internship but are generally more focused on course work than original research or thesis degrees. Sometimes these professional degrees are also abbreviated to MS or MA, but they can have many different names, such as MF for Master of Forestry or MESM for Master of Environmental Science and Management. Even the MBA and JD degrees linked to business and law schools, respectively, might be the right choice for you. Although most master's students do not go on to a career in research, thesis-based master's degrees are still common in the conservation field. It is widely believed that designing, executing, and presenting a substantial research project is an invaluable experience for aspiring professionals, in part because it will prepare you to assess research with a critical eye. That said, professional degrees without a thesis are the best choice for many students, particularly those who still need the background provided by substantial course work. This is commonly true of people who are shifting into conservation from a different career track (e.g., an aeronautical engineer) or those who had a very broad-based undergraduate degree (e.g., a liberal arts student who took as many courses in literature and history as biology or environmental policy).

Can you climb the academic tree too far and find that you are overqualified for your preferred career and that people will not hire you because they assume you will soon leave for something better? There is some risk of this but it is fairly minimal. People with PhD degrees now hold a wide variety of jobs in conservation, not just in research institutions and universities.

> *I have a friend with a doctorate in genetics from Harvard who worked for years as a state conservation officer, a position in which she focused on finding poachers rather than describing genomes.*

Certainly it is best to err on the side of being a bit overqualified for a job than to be a bit underqualified.

The other side of "what degree?" is the subject area, and with conservation the list of titles is very long: environmental science, conservation, oceanography, wildlife, forestry, policy, anthropology, biology, sociology, landscape architecture, economics, planning, law, and so on, dozens of possibilities in all. Therefore, the actual name of your degree is not enormously important when you go searching for a job; most prospective employers will look beyond it to examine your work and project experience and the courses you took. Occasionally you may hit a snag with a large government bureaucracy that decrees that for job X you must have a certain type of degree, but even in these situations it is usually not difficult to explain how your education has prepared you for a given post.

The question of a degree name may well be a significant issue with respect to the course work you must take to achieve the degree. If you are working on the effects of forest harvesting on biodiversity under the banner of a forestry degree, you will probably be expected to learn much more basic forestry than if you were pursuing the same topic under a natural resources degree. Certainly you will want to have the latitude to choose any courses that you need to broaden your skills as a conservation professional (e.g., social-science classes for the biologists and earth scientists and vice versa for the social scientists). Flexibility about course work varies greatly among different universities. At some universities it even extends to allowing you to work on the same topic with the same advisor for two or three alternative degrees.

Advisor

Although you may aspire to work on subject X, or to enroll at university Y, rather than to study with Professor Z, selecting an advisor may be the single most important element in deciding where to apply. For professional degree programs you may not be able to choose an advisor until you are enrolled, but for a degree that involves thesis research this will probably be a key part of the selection process. A good advisor may become a lifelong colleague, mentor, and inspiration, and open doors to a network of colleagues and future opportunities.

Selecting an advisor may be the most important part of your decision.

A bad or incompatible advisor may doom you to years of misery and perhaps failure. If you have identified a topical area, identifying a list of potential advisors is straightforward: they are the people who have contributed to the literature or conservation advancement within your topic of interest. Similarly, if you have a shortlist of universities, they are the relevant faculty of your targeted universities. There are three basic ways to make a preliminary evaluation of potential advisors: reading their websites, talking to people who know them, and reading the papers, books, and conservation outputs that they have produced. Reading websites is clearly a first step and will probably filter out some names quickly. Do not be swayed by elaborate websites; many of the best potential advisors are too busy doing their job to worry about how they look in cyberspace. Selecting an advisor is probably the most important part of your decision.

Next, try to find some people you can interview. The degrees of separation in conservation circles are very short and you probably already have contact with some faculty and students who know, either directly or by reputation, a potential advisor. At this stage, two broad questions will probably suffice: (1) does this person have a reputation as a good graduate student advisor?; and (2) is this person known for producing high-quality work? Ideally you will want someone who is both a good advisor and well known and admired in the field, but they do not always go hand in hand and the former is more important to you than the latter. The less tangible qualities of potential advisors will become more apparent when you meet them.

Lastly, you should read some publications produced by potential advisors and their students. Most faculty members' websites will provide a partial list of publications and to complete this and make sure it is up-to-date you can use an online search engine. You do not have to read every paper thoroughly. A modest sample, focusing on the abstract, introduction, and discussion should give you a good taste for a person's work. Having an advisor who has a strong publication record will be of particular importance if you plan to pursue conservation research, but less so for many other aspects of conservation work.

With all this information to hand you will be able to decide who you want to actually contact, a process described in Chapter 4. Hang on to the publications because they will be useful when you are interviewing and ultimately selecting an advisor.

WHEN TO BEGIN?

As implied in Chapter 2, it is wise to be thinking about a graduate degree throughout your undergraduate career, but this process will need to move into high gear starting early in your final year if you want to go directly into a graduate program. Even if you plan to take a year or two off to strengthen your work experience, it would be a good idea to start the selection process while you are based at a university where you can easily solicit advice.

One of the more important pieces of advice to obtain is the best time of year to begin your search in your country. Many graduate programs begin in the autumn, along with the regular academic year, and it is likely that these positions were filled several months earlier, during the preceding winter or spring, but sometimes selection can begin almost a year in advance. Some positions may be filled at almost any time of year because occasionally graduate students start on a schedule dictated by field work or funding rather than courses.

SOME SCENARIOS

To demonstrate how these four issues – university, topic, degree, and advisor – come together while choosing where to apply, we will describe some of the most common scenarios for finding a graduate program and some variations on these. These scenarios are most applicable to research-based degrees. For a professional degree, the process is more analogous to applying for a bachelor's degree program, particularly because you may be responsible for most or all of the funding for your program.

The complete-package scenario

One simple scenario combines the key elements into a single package that allows you to pursue a topic, advisor, university, and degree all in one integrated process. Here is how it works. Dr Black obtains funds to support a graduate student to undertake research on a particular issue, for example evaluating public attitudes about a plan to reintroduce an endangered snake species, and she advertises this opportunity on one or more websites[3.2] with instructions on how to apply. If the position sounds of interest and you meet the qualifications listed then you just need to follow a set path of application; topic, advisor, and university are all preordained. Usually the degree level is also specified as master's or doctoral, typically based on the number of years of funding available and the intellectual scope of the project. If a project seems perfect but is the wrong degree level for you, it may be worth inquiring anyway because there might be some flexibility on this issue. For example, the advisor might be willing to seek additional funding to expand a master's project into a doctoral-level one in order to attract an excellent candidate. In one variation of this scenario, faculty members do not bother to advertise an opportunity because they have so many general inquiries from prospective students that they are confident that they can always find a strong candidate. This is particularly likely to be true if they work on a high-profile topic, like conservation of marine mammals or large carnivores.

The open-ended scenario

Under this scenario you will apply to a university program, say the Department of Environmental Sciences at Alpha University, where your application will be reviewed by a committee. If they accept you, there is a good chance that you will also be allocated a university-funded financial package that includes both your tuition costs and a stipend of some type. At North American universities, the stipend is often based on a teaching assistantship that will require 10–30 hours per week during the semesters (great experience but a lot of work). Sometimes your assistantship will require you to work on someone else's research while you develop your own project separately. Mixed packages are also common; you may start on a teaching assistantship then move to financial support provided through your project. Probably the best deal is a financial package, often called a fellowship or scholarship, which has no particular work assignment; you just have to complete the requirements of your program. If you are accepted without an offer of financial support then you are effectively placed in one of the financially independent scenarios described below.

We call this the open-ended scenario because your advisor and research topic are unidentified at the time of application and acceptance. This may be resolved soon after acceptance or it may continue for months. Many open-ended students complete their first semester or two working with a temporary advisor while courting a long-term advisor with whom they will develop a thesis topic. If your interests are quite broad, this scenario has a real advantage because after a few weeks or months in a department you will have a pretty good idea of who would make a compatible advisor and a better sense of where your project interests are headed. However, there is a real risk that you may find yourself in a department with no compatible advisor, at least one who is willing and able to work with you, and thus you may find yourself stretching the boundaries of your area of interest. For example, you might go to Alpha University planning to work on soil erosion and discover that of your two potential advisors, Dr Brown is a jerk and Dr Green is not taking on any new students for 2 years. Perhaps you will end up studying soil microbiology with Dr White … and if you are lucky you may discover that you love microbiology. The bottom line is that the narrower your interests, the greater the risks associated with the open-ended scenario.

The stacked-deck scenario

This is a variation on the open-ended scenario in which you try to reduce the uncertainty by identifying a potential advisor early in the process. Ideally this will happen long before you formally apply. If your application package contains a note from Dr Gray stating that he has communicated with you and would welcome the opportunity to work with you if you are accepted into the program, then this will probably carry significant weight with the selection committee. Another possibility is to identify Dr Gray and develop a relationship between the time you are made an offer and the time you have to decide whether or not to take up the offer. Unfortunately this period is usually too short for this to work well.

The difficulty with the stacked-deck scenario is that it assumes you can identify Dr Gray and get his attention. Unlike Dr Black, who must actively find a prospective student to undertake her project in the complete-package scenario, Dr Gray may be keeping a low profile. He might only be scanning his emails for new students with one eye open, perhaps because he is half asleep, perhaps because he is fully occupied keeping his eye on a large set of students already. You may be able to get through to Dr Gray using the approaches described in the next chapter, or you may be able to increase your chances by meeting him face-to-face, perhaps at a regional conference, or through an introduction by your advisor or someone else in your social network.

The financially independent scenarios

If you are one of the lucky few people who does not need any university-based financial support for your education, this can change the playing field dramatically. It opens up many opportunities for professional degrees (which are usually self-funded) and for research degrees it will give you more latitude in finding an advisor and selecting a project. Nevertheless, it does not mean the door is wide open because you will still need to meet certain standards. Fortunately for you, it is usually much easier to meet threshold standards than to compete with scores of candidates for a handful of university assistantships. There are three major ways a prospective student may be financially independent, which are sort of sub-scenarios.

The agency-support scenario applies primarily to employees of conservation agencies who are allowed to set aside, at least in part, their regular duties in order to pursue a degree, often focusing on a topic of particular interest to the agency. Typically the agency will continue to pay your salary and may cover any project expenses; tuition may come from the agency, the university, or you. The agency gets a better-educated employee and often a useful project; the employee gets career advancement; and the advisor and university get a motivated, experienced student. Geography often constrains this scenario; if there is not a suitable advisor and university near the work site it is more difficult to make it work.

Under the prestigious-fellowship scenario you need to compete against hundreds, perhaps thousands, of other students across your nation (or even the world) for a fellowship that will cover your tuition and stipend, and sometimes research expenses too. Some of these are sponsored by private foundations; most are funded by government research agencies. Someone on your campus will have a list of those available in your country. These competitions are usually so rigorous that the winners will find that most (but by no means all) potential advisors are willing to work with them. If you are lucky enough to be one of these awardees you may be in a position to bargain for substantial support for your research work. For example, you could compete for one of the complete packages described above and propose using your stipend and tuition funds to allow more funds to be directed toward research expenses.

The personal-finances scenario covers students whose personal wealth, whether given by Mom and Dad, or earned through a successful career in another field, allows them to cover their own tuition and living expenses. Some universities have large numbers of these students, especially earning professional degrees without a thesis. Personally funded students are generally uncommon in thesis-based graduate programs where the research-project expenses may exceed the cost of tuition and living costs.

The independently initiated project scenario

Are you attracted to pursuing a graduate degree because you really want to undertake a particular project that has captured your interest? You may even have most of the details worked out – goals and objectives, design and methods, end products, etc. – and all you need is an advisor and some funding. If you have investigated this topic in depth you will probably know who the potential advisors for this project are, and the next step is convincing one of them to accept you as an advisee. There is a reasonable chance of this happening if: (1) a faculty member is impressed by the creativity and feasibility of your project idea; (2) they have the time to take on a new student; and (3) they can see a clear path to funding for the project (perhaps via the stacked-deck scenario). The riskier your proposed project is (e.g., uncertain outcomes, long periods far from the university, difficult funding) the harder it will be to find an advisor willing to take the risk. However, if you have a sound idea and are passionate and persistent about it, this independent approach can pay off.

Hybrid scenarios

Naturally, these scenarios do not capture all the variation that is out there. Some universities have well-established systems for selecting students that do not fit well into any of these alternatives, and at every university there will be some students who are mavericks who follow an unusual path into

Many paths can lead to a great educational opportunity, so explore a diversity of options.

the program. In particular you may have to work with a mixture of financial support generated by you and the university. If you keep these dominant patterns in mind and remain alert to any variations, then you will find your way down the path without getting badly lost.

Three real-world examples

When I have funds for a new graduate student I start by advertising on three websites, our department's, and the job boards of the Society for Conservation Biology and the Texas A&M University Wildlife and Fisheries Science Department. I ask for letters of application and résumés plus scans of transcripts and GRE scores. Typically I will get about 30 applications and from these I will select the top 5 to 10 candidates and ask them for a sample of their professional writing and letters of recommendation. Next I go to phone interviews with between four and six people and generate a ranked list of my top candidates. In the final stage I invite my top candidate to come for a visit and usually by the end of a couple days I am comfortable making an offer. It is not uncommon for my top one or two candidates to pass on the offer, often because they have partners who are worried about job opportunities in a small town.

When students contact me about the possibility of starting a graduate program, I try to be as encouraging as possible, but I explain that the key issue is that they must be successful in getting a scholarship from our university or the government. I ask the student to send me a CV and to outline some of their broad areas of interest. If they seem likely to get a scholarship, then it's sensible to continue correspondence; it would be unfair to continue to encourage someone whose chances are limited. I usually list some options for possible projects and what additional financial and logistical support may be available for them. These e-mail exchanges will often be intersected by phone calls and/or visits to discuss potential projects in person. During these exchanges it's usually

> easy to work out if they have the enthusiasm and drive to successfully complete a graduate program. If it seems very likely that the student will go ahead with the project, I will often take them to potential field sites and help develop ideas further for the project proposal that they need to write as part of the scholarship application.

> Each year, I have positions for students provided by our national research funding agency but I never advertise for candidates. Instead, I wait for prospective candidates to contact me. I then assess them primarily on their levels of enthusiasm and passion – key indicators of a successful mentor–student relationship. Most importantly, I must like them and they me. We then start the sometimes difficult journey of learning about and from each other in an environment that is devoid of fear and emotional hierarchy. Indeed, I think the emotional aspects of the mentor–student relationship are more important than the intellectual ones.

This last anecdote is a good bridge to our next chapter, where we will discuss the process of choosing and being chosen in some depth.

Further readings and notes

3.1 These websites can be found by searching the Internet directly. A large database on universities with conservation degree programs is maintained at www.conbio.org/professional-development/academic-programs and for a broader approach you can start with GradSchools.com.

3.2 Websites that list specific openings for graduate students are typically the same sites that list conservation jobs (see Chapter 9), although relatively few appear at sites where it is expensive to list an opening (notably in the journals *Nature* and *Science* and their associated websites). In addition, most university departments will list openings on their own websites.

3.3 To link a topical area to a university or location in Web of Knowledge, you could type, for example, "amphibian AND disease" in the Topic field and "Sydney OR

Yale OR Idaho" in the Address field to produce a short list of amphibian disease papers that have one of these three locations in the Address field for at least one author. Note that "Sydney" will generate papers from both the University of Sydney and other institutions in Sydney. If you are interested in all the institutions in a particular country or state you can use that in the address field, but be aware that many abbreviations and alternative name and spellings are used (e.g., UK/England, Brazil/Brasil). Thus "Idaho" will lead to University of Idaho or Idaho State University, but if you were interested in all the universities in the state of Idaho you should also try "ID." See further instructions on Web of Knowledge.

4 Applying for admission

After a period spent mapping the landscape of possibilities, you are ready to begin the actual process of applying. If you have done a good job with your investigation, keeping ample notes on all the viable alternatives, then this process can be quite efficient. While you are applying, the landscape may change – in particular, new openings may be advertised – so be sure to keep checking relevant websites.

Your first steps may go in several directions, especially if you pursue multiple, different scenarios. At Alpha and Beta Universities you may be applying for admission as per the open-ended scenario and you only need to carefully follow the steps of their formal application process. At the same time you might want to pursue the position that Dr Black has advertised at Gamma University and make general inquiries with a whole rainbow of potential advisors at various universities who share your interests, including some at Alpha and Beta Universities. And in your spare time you can try for a couple of those prestigious fellowships if you have excellent credentials.

The following sections are not necessarily in chronological order because first steps vary among scenarios, but you will probably hit all of these steps along the way.

MAKING CONTACT

In the best of all possible worlds, you will be in direct communication with several potential advisors early in the process. Initiating communication is easy if a faculty member has advertised a position; they will be searching for good

Saving the Earth as a Career: Advice on Becoming a Conservation Professional, Second Edition. Malcolm L. Hunter, Jr., David B. Lindenmayer, and Aram J. K. Calhoun.

students just as you are on the prowl for a good advisor. For these folks, your first contact can be your letter of application, which we will cover in the next section. More commonly you will be making what salespeople refer to as a cold call, contacting someone whose work interests you, but with no idea if they are open to accepting a new graduate student. Some of these people may have openings that they have not advertised; some may be candidates for the stacked-deck scenario or one of the independent scenarios described in the preceding chapter.

For cold calling we suggest a short e-mail that covers just two basic points. First, you are looking for an advisor who shares your interest in climate-change policy or regional planning or whatever. The closer the match between your interests and the advisor the better, but be careful not to restrict yourself too much. You can minimize this risk by using language such as "I am especially interested in the effects of overharvesting on intertidal communities, but would be happy to consider any opportunity in marine conservation."

Second, you need to succinctly (in one meaty paragraph) convince your reader that you are a viable candidate for a student opening. Naturally that will include your current status ("I am in my final year as an environmental policy student at Omega University") and some key points from your résumé ("I have undertaken an honor's thesis on ... served as President of ... "). Be sure to emphasize your relevant work experience ("I have spent my last three summers working on three different projects ... "). If you have strong numbers from your grades or standardized tests (e.g., GRE or TOEFL) it does not hurt to mention these too. Finally, you should attach a résumé so more detailed information is readily available, but does not clutter up the e-mail itself. If you are pursuing the independently-initiated project scenario described in the preceding chapter you would also include a short proposal, perhaps 1,000–4,000 words.

Do not use these initial inquiry e-mails to ask broad questions, such as "How does the application process work at Alpha University?" or "Could you please tell me more about your research program?" Even for specific questions – for example, "Is there an application deadline?" – you should first look very hard to find the answer yourself on the relevant website or through contacting the university's graduate studies office, and then you should save any unanswered questions for a follow-up e-mail if your first is answered positively.

You may send these inquiry e-mails to a few dozen people and you will get many "Sorry's" and many who do not respond, but you can be reasonably confident that you will hear from anyone who has an opening for which you are qualified. There is a small danger of having your e-mail being discarded without being read. For that reason it is best to be explicit in the subject line (e.g., Inquiry about graduate study) and to avoid sending e-mails during periods (summers and holiday periods) when the faculty member may quickly purge a large backlog of e-mails. If you have no response after a month, you could send a second e-mail, especially if your first round did not generate many responses.

You can also increase your chances of receiving a response if you personalize your e-mail for a given faculty member. No music sounds so sweet to an academic's ear as, "I have read your paper on land use policy and ... " or better yet, " ... two of my summer jobs were related to your work on ... " It may not be practical to do this for many faculty members (yes, you do have to read that paper on land use policy), but if you have a few high-priority targets then this is a worthwhile investment.

Conversely, avoid anything that would suggest that you are sending this e-mail to hundreds of potential advisors, such as using "Dear Professor" as the salutation rather than "Dear Dr Gray." It is just as bad to send the same e-mail with multiple faculty members in the To: field. This might seem like a reasonable thing to do if you are writing to several faculty members in the same program that uses an open-ended scenario, but in this case, send individually addressed e-mails sequentially with a short interval (perhaps a day or two) between each, and mention that you are contacting other faculty in the department.

Your initial inquiry should succinctly describe who you are and what you are seeking without any spelling or grammatical errors; think of this as making a "first impression" through written communication.

Finally, and very importantly, do not make mistakes in spelling or grammar at this stage or any other, and don't trust word-processing software to catch all errors. Most faculty members are careful, detail-oriented people who value proper communication and that does not extend to the conventions that are acceptable for text-messaging. Everyone makes mistakes occasionally, but if you make even one small mistake in a simple letter it warns the faculty member that you are likely to make hundreds of mistakes in a thesis-length document and that you are not taking the application process very seriously. In summary, this inquiry is your opportunity to make a "first impression" and it may well determine whether a face-to-face "first impression" ever occurs.

PERSONAL ESSAY OR LETTER

Sooner or later you are going to have to demonstrate that you are an ambitious worker with a clear vision of your future, a creative thinker who understands your field of interest and wants to advance its cutting edge, and a friendly person who works well with others. As a subtext you will need to make the case that a relationship between you, Dr Black, and Gamma University will be mutually beneficial to all parties. And all of this has to fit in a page or two of eloquent prose. Yes, this is a daunting task. Start early, write and rewrite it, share it with a couple of critical readers, and rewrite it again, and again. This essay is more important than any term paper you ever wrote so writing four or more drafts is well worth the effort.

This document may distill some key elements of your résumé but it is not just a repackaging of all those dry facts. It will not be scrutinized for data, but rather to figure out who is the real you. Thus it needs to be scrupulously honest and frank. It is fine to let your unique personality shine through (in this sense it is very different from scientific writing), and to indulge in a bit of introspection, especially as it relates to your academic and career interests. You want to seem like an interesting person, but of course you do not want to come off as weird or overly eccentric. Let your unique personality shine through an application letter.

Let your unique personality shine through an application letter.

Secondarily, but still importantly, the letter will be examined to see if you are a capable writer. If you have written major papers such as an honor's thesis, and especially if you have published a paper, you should offer to send these along too and they will also be evaluated to judge your writing ability.

This personal document may take the form of a letter to Dr Black in response to her advertisement, or to Dr Gray who responded to your initial inquiry. Often it is an essay that is required as part of a formal application. It may start as a letter to a particular faculty member and later, if you are given a green light for formal application, need to be recrafted as an essay or personal statement detailing your academic interests. Although it may have to be tailored to each application, over 95% of it can probably be used multiple times. Be especially careful about the 5% that you change; it is easy to mistakes while cutting, copying, and pasting text.

The bottom line is this: take the time to do it right, including paying attention to those pesky spelling and grammar issues, and crafting the letter for each distinct position.

INITIAL CONVERSATIONS

Most people believe they can judge the character of a person to a significant extent by the way they speak (not withstanding all the successful actors and salespeople in the world). Thus, your potential advisor is likely to want to talk with you *if* you are a prime candidate for an opening. Generally this is a situation where you should probably wait to be spoken to. You can gently suggest making a date for a phone call or videoconference, but are likely to get a lukewarm response unless the process has unfolded to a point where the advisor knows you are one of the top candidates. If you are in the enviable position of being able to choose among multiple offers, then you may want to push for a conversation to understand your choices better; but again this will probably come relatively late in the process.

Conversations are often a better medium than e-mails for asking and answering questions about the details of a project or the mechanics of an application process. Indeed, you should have a set of questions prepared ahead of time. However, be sure to do your homework first to avoid simple questions on application logistics that you could easily answer on your own. That old aphorism about "The only dumb question is the one that is not asked" is a device to catalyze class discussion: it does not apply to questions derived from laziness. You should also try to anticipate likely questions so that you are better prepared to answer thoughtfully "on your feet."

APPLICATION MECHANICS

At some point you will have to fill out forms, submit transcripts and standardized test scores, pay application fees, and so on to complete the formal application processes for each university to which you apply. Needless to say, these have to be done carefully, both to avoid pitfalls later and to avoid giving the impression of being sloppy. Detailed instructions will be available on the university's website.

If you are working through an individual faculty member for a specific opening, you may or may not have to go through the formal application early in your interactions. Some faculty members will want to see all the elements of a formal application from the beginning; others will spare you this expense and work until you are far along in the selection process, perhaps even after you have been offered a position. They may be satisfied, at least initially, with your letter, résumé, references, and perhaps scans of unofficial transcripts and test scores. Ultimately, however, you will have to go through the formal application process under any scenario to be admitted into a graduate degree program.

Beware of deadlines. Sometimes applying for graduate school is quite open-ended (especially under the complete-package scenario), but often there are strict deadlines, especially if you wish to be eligible for university-sponsored financial support. At some universities, deadlines are surprisingly early, nearly a year before your intended starting date.

REFERENCES

Input from some people who know you will be required at some stage and, as mentioned in Chapter 2, it is best to anticipate this by cultivating some faculty members and employers to serve as references months or years before you need them. You want people who have witnessed a broad set of your sterling qualities. You may only need to provide their names, positions, and contact information (regular address, e-mail address, and phone number) and your potential advisor may contact them to learn more about you. More commonly you will need to ask them to write a letter for you that will go to either your potential advisor or into the formal application, or both. They may also need to fill out a specific form for the formal graduate application.

In all cases, it is important to make this an easy process by giving them everything they need in a packet: (1) name and address for their letter; (2) instructions about mechanics (e.g., e-mail versus website submission, and any required forms); (3) copies of your résumé and application essay; and (4) a link to any specific information about the opening. The résumé and essay will give them good fodder for their letter and creates an opportunity to ask for their feedback. Once a person has written one letter for you, it is relatively easy for them to cut and paste new names and addresses and make other minor adjustments to send the letter off multiple times. That said, you definitely risk "reference fatigue" if you ask for dozens of letters, so this is another good reason to evaluate carefully where to apply. Furthermore, it is often better not to ask for letters for a particular position until you have reached a stage where they are needed. For example, some faculty members will only want to see names of references initially, and they will ask for actual letters only after they have selected some finalists.

VISITING

Have you ever been on a blind date? Would you be interested in going on a blind date that lasted for a few years? Choosing a program without a visit to see the university and meet your advisor and future colleagues is like going on a very long, very important blind date. Of course, it may cost a significant sum of money for a visit, but this is cheap insurance compared to the risk of spending years of your life

Choosing a program without a visit to see the university and meet your advisor and future colleagues is like going on a very long, very important blind date.

in the wrong place. It is possible that your prospective advisor will be willing to pay for the whole trip or a portion of it (e.g., providing local room and board) if you are a top finalist for a funded position. Thus it is worth gently asking for financial assistance for a visit, but do not be put off if your request is refused. Your prospective advisor may have numerous experiences with finalists who came for a visit, were made an offer, and then went elsewhere.

The risk of making a trip that does not lead to a position will be minimized if you plan to visit late in the selection process, after you have been identified as one of the leading candidates for a position, and after you have a clear idea of your preferences. Visiting universities to filter a long list of places that interest you is generally a waste of time and money. You will also get much more attention from a prospective advisor during the final stages of selection. A student who is just traveling around, searching for possible openings, is lucky to get more than half an hour of a faculty member's time; a finalist for an

opening is likely to get a couple days with multiple meetings, meals, sessions with current students, and tours of campus and field sites. After that kind of interaction you should know whether you are ready for a commitment that will last for years.

Two alternatives to visiting your prospective advisor on campus merit mention: visiting campus while your advisor is away (perhaps on sabbatical in another country) or visiting your advisor away from campus (perhaps at a professional conference). Both are less than ideal because your ability to gather information from multiple sources will be constrained, but better than no visit at all.

FIRST IMPRESSIONS

Although you have been trying to make a good impression on your prospective advisor since your first e-mail, this process will move to a whole new plane with your first meeting. Do you have a navel pin shaped like an orchid? A tattoo of a cobra on your arm? It is best to keep them covered up at your

You never get a second chance to make a first impression.

initial meeting. Even though your prospective advisor undoubtedly shares your love for biodiversity, a shirt or blouse that you might wear to work in an office (sometimes called smart casual) will be more appropriate than a t-shirt covered in coral reef biota. Academics tend to be quite liberal about personal appearances, but it is still best to err on the conservative side in the beginning.

At least as important as your visual appearance is the way you speak. Academics spend much of their time thinking about language – editing papers, speaking to classes, etc. – and they will not be impressed by unthinking use of the jargon of the day: "Oh my God, this place is totally awesome." On the other hand, you don't want to sound like a nerd who walked out of the *Oxford English Dictionary*: "I have been favorably impressed by the diverse amenities that are available here." Simply sounding like an intelligent, thoughtful, interesting, articulate person is all that is necessary. Okay, that may seem like quite a challenge, but you get the main idea: using contemporary argot (is that an interesting word or a nerdy one?) to show how cool you are is not the goal.

It is also important to convey that you are a reliable person, and that means being on time for appointments. Arrive on campus early to give yourself a buffer. As the old saying goes, you never get a second chance to make a first impression.

Finally, first impressions can come long before you visit campus. How you sound and look during a videoconference – even the background visible behind you – create an impression. In this age of Internet culture, you probably also have made some kind of impression in cyberspace (e.g., blog spaces, personal websites, or even employment directories) or within online social media spheres (e.g., Twitter, Facebook, Instagram, or others). If your cyberspace signature is accessible to prospective advisors, there is a good chance that some of them will have a look.

INTERVIEWING A PROSPECTIVE ADVISOR

Eventually the table will turn and you will be interviewing your advisor. Your life does not depend on being accepted at Gamma University. Perhaps you will go to Sigma University, perhaps you will pursue another dream, becoming a professional athlete or traveling the globe for a couple years' bird-watching, scuba-diving, and volunteering on conservation projects. To decide whether this is the right opportunity for you, it will be helpful to talk to your potential advisor in person. The chemistry between an advisor and advisee has to be right and this is much easier to judge in person.

Beyond the subjectivity of compatible personalities, there are some specific issues you should explore, starting with an advisor's approach to working with graduate students. Some give their students free rein to design and execute their own projects; others like to work closely with their students, for example, visiting field sites frequently and pitching in to the task at hand. Accessibility is an issue; will you be expected to meet with your advisor once or twice a week on a regular schedule, or will you have to initiate any meetings and make an appointment long in advance? Will you be your advisor's only student over the next few years or one of a dozen? Will you be one of their first graduate students or one of their last? There are no right or wrong answers here; just a context and style of working that may or may not be right for you. Asking someone what they see as their strongest and weakest attributes as an advisor might elicit some interesting introspection.

Naturally, if there is an independent project required this will be a major topic of conversation. Are there ample opportunities to pursue exciting new questions, or are the constraints of time, resources, and funding-agency expectations likely to keep you tightly reined? Is there much risk involved; for example, can you easily collect your data with minimal chance of failure? How does this particular project fit into your advisor's overall research program? Is this a short-term area of investigation or one that is likely to continue for years? Are there other collaborators who you should meet or call to discuss their views of the project? What are the expectations with respect to publishing papers and authorship? You particularly need to develop a good understanding of how a project will fit with other responsibilities. Will you be expected to teach and work on other projects besides your own? With your particular background what sort of course load is likely?

It is important to have a clear understanding of the financial picture. Usually the amount of any stipend will be specified and is probably not negotiable, but you should have a clear idea of the duration of stipend support. This issue has two aspects. First, you need to know how many years of funding are guaranteed by the university or some other institution. Hopefully all the funding is in place, but it is not uncommon for an advisor to recruit a graduate student with, for example, only 2 years of stipend support in hand and promising leads for a final year. Obviously this is a bit risky because that third year may not materialize, but many departments have a very good record of stepping in under these circumstances and supporting a student through to completion. It is devastating for all parties to have a student leave for financial reasons. Second, you need to know when a stipend will terminate with or without available funds. At many universities you will be expected to finish your work in 2, 3, or 4 years whether or not project funding is still available, and if you go over that time limit you will have to continue without a stipend. What is normal differs greatly among universities, as does the frequency of exceptions to the norm, so it is important to clarify all this before you make a commitment. The situation also differs between the social and natural sciences: to have funding in advance for your whole program is less common in the social sciences.

Similarly, you need to have a solid understanding of the financial support that may be available for any project work. How much has been secured? How much is covered by pending proposals? How much, if any, are you expected to find? Is there in-kind support (e.g., vehicles and field housing provided by a government agency)? How much support is available for conference travel? Hopefully you will be able to present your work at a major conference toward the end of your tenure. Other university resources should be discussed, such as office space, computer hardware and software, information technology support, and library resources.

The discussion above has focused on the nuts and bolts of working with an advisor and these are important issues, but ultimately they may be trumped by issues that are less tangible: the chemistry of a relationship.

> *My experience has led me to believe that a good advisor is one who is willing to allow graduate students to nourish their own curiosity and who is willing to establish a relationship of mutual learning. In addition to assessing your advisor's intellectual intelligence, be careful to consider whether he or she will support your independent growth, both intellectually and emotionally.*

INTERVIEWING OTHER STUDENTS

Without question the best authorities on a person's attributes as a graduate advisor are the people who have been down this path or are walking it now.

The selection process is a two-way street and you need to interview prospective advisors and fellow students carefully.

Former students will have the longest experiences, and will have had some time to let their experiences gel. If there are some nearby, go and visit them; if not, e-mail them to make an appointment for a phone call or videochat. (Your prospective advisor should have contact information readily available.)

The best time to interview current students is during a campus visit. If you go to lunch with a group of students, this is an ideal opportunity to ask about aspects of the graduate program in general: do you have a good seminar series? Do students help one another with their projects and is there a culture of mutual learning? Do faculty members share their resources and expertise with all the graduate students in the department or just their own? How is the camaraderie among students and between students and faculty? Are there informal gatherings to foster social and professional relationships? Beyond asking questions, you can also learn a lot about a group of people by just observing how they interact. Some logistical issues are best addressed by students; for example, how much does housing cost in the area, and is it easy to get around without a car? A university's stipend might look a quite a bit larger or smaller depending on how the local cost of living differs from where you are currently living. You should ask the students for their opinions of the adequacy of departmental and university resources such as office space, computers, and the library because these may differ from your prospective advisor's view.

For personal questions about a particular advisor, it is better to talk to people one on one. Some people may be less forthcoming in a group for fear that they are being a tattletale. The more time you spend with someone the more frank they will be, another good reason for visiting a campus rather than trying to have phone conversations. Generic questions, such as what are the key things you like and dislike about Dr Black, are one way to begin. Basic logistical questions (as outlined above) can also be good "ice-breakers" leading to more nuanced questions. Ultimately you should be asking about working style: is Dr Black a micro-manager always looking over the shoulder of her students or off on another planet and difficult to reach? (Of course, you will want to avoid leading language like this. It is important to use neutral wording.) When you give Dr Black a proposal or chapter to review, how long does it usually take to get it back? If you need 10 minutes to discuss an issue, do you usually need to make an appointment? Be prepared to hear that most graduate students do not think their advisor spends enough time with them. At some level, almost everybody's worldview revolves around themselves, and this can easily translate into feeling a bit neglected by key colleagues. Nevertheless, you should be especially alert to overt signs of neglect by advisors expressed by multiple sources.

> *I recall a fellow student complaining about her advisor. The only contact she had came when they passed each other in the hallway every few months and the advisor shouted, "Hope everything is alright!"*

On the other side of the coin, quality counts for more than quantity when it comes to advice. It's probably better to have a thoughtful and efficient advisor you see for minutes per week, than one who spends hours per week giving you useless advice. Similarly, it may well be better to have an advisor who spends more time in the office minding the administrative side of the project than in the field helping you feed mosquitoes or conduct repetitive social surveys.

Although the advisees of a given faculty member are going to be your best sources, it is also worth talking to other students, especially in smaller programs where everyone knows everyone else. If the graduate students of a particular advisor routinely have problems, these other students might be more likely to talk about it than the advisees themselves and may offer greater objectivity. Other graduate students in the program may also have insights about departmental culture with regard to financial support for students whose regular funding evaporates.

Whether you hear more praise or criticism, ask people to be as specific as possible. This will help you judge the reliability of these sources. And of course, try to get a large sample size by talking to many people, hopefully both new and more senior students.

INTERVIEWING OTHER FACULTY

It is also worth talking to some other faculty members. Tell them a bit about yourself and ask them about the graduate courses they teach and their research programs. Ask them if they have served on any advisory committees for other

Interview other faculty members beside your prospective advisor.

students of your advisor. All this will help you decide if this is a congenial environment, and it might be the first step in selecting people to serve on

your advisory committee. It is not uncommon for faculty to ask one another for second opinions: "What did you think of that student I sent to talk with you this morning?" Indeed, some faculty members also make a point of asking their graduate students to evaluate prospective students, so be aware that you may be "on stage" during your entire visit.

If the funding for your project is incomplete, you should be sure to talk with the chair of the department or the graduate program coordinator to ask about the department's track record of helping students to find funding. Even if all of your funding appears to be in place, it is a good issue to raise because bad things happen; for example, an agency might renege on a supposedly firm commitment.

MAKING A DECISION

If you are lucky, at the end of this process one clear choice will evolve and you will be offered that position. Life is not always so simple, of course. What happens, for example, if you have a top choice that will not be decided upon for another month, but this morning you were made an offer for your second choice? There are no standard procedures for getting through such dilemmas, just one maxim: be open and honest. Here is a truncated dialogue to illustrate how this might unfold.

Dr Black: "Good news Pat; I am calling to offer you the position."

Pat: "That's wonderful news; thank you. I am also a finalist at Beta University, but unfortunately I won't hear about that until the end of the month."

Dr Black:	"Hmmm; that's awkward because I have a couple other good candidates and I can't wait that long or they'll take up other offers."
Pat:	"Okay, let me talk to the people at Beta and I will get back to you soon."
Dr Black:	"I can wait until next Monday and then I'll have to make the offer to someone else."
Pat:	"Dr Gray, this is Pat. I've just received an offer at Gamma University. I told them that I was a finalist with you but would not know the outcome until the end of the month. They can't wait that long so I have to make a decision by next Monday."
Dr Gray:	"Congratulations Pat. I really hate to lose you as a candidate but there's no way to resolve this by next Monday because I'm interviewing another candidate next Wednesday."
Pat:	"Okay, I'll have to think hard about this. My first choice would be to work with you but I may have to take the bird-in-the-hand."
Dr Gray:	"I totally understand. Just let me know your decision as soon as you've made it."

So what will Pat do? Of course, there is no right answer. In theory one could devise a quantitative solution by estimating the odds of success at Beta University and using some system to quantitatively compare the opportunities at Beta and Gamma Universities. In practice, some subjectivity will shape your thinking about all the factors that have been discussed above – your potential advisor, funding availability, other students, the reputation of the department – as well as many important factors that have not been covered, such as the quality of life. Is this a part of the world where you would enjoy living for a few years? Do you have a spouse or partner or children? If so, then their needs are going to be paramount too and you should try to have them participate in your site visit. After making a decision like this, one that will likely impact the rest of your life, choosing between pizza or baked beans for dinner will seem very simple indeed.

5 Designing a program of study

So your success as an undergraduate has earned you a place in a graduate program — congratulations! — but now the rules of the game are about to change. Earning high grades and undertaking some relevant jobs may have brought you here, but you will need additional skills to go forward with continued success. As you approach fledging into a career, you really need to develop the talents that will make you a competent and effective conservation professional and this will require that you design your program with considerable forethought.[5.1]

YOUR GOAL

It is important to be clear that earning a degree is not your ultimate goal; it is merely a means to an end, a journey toward your rewarding career as a conservation professional. Consequently, as you undertake the requirements for your degree you should always be asking yourself, how can I complete this task in a manner that will prepare me for my career? This issue should be a dominant theme for one of your first conversations with your advisor and anyone else who joins your advisory committee. As professional academics, their eyes are probably focused on a closer target, finishing your degree in a timely and laudable manner, so you may need to remind them occasionally of your ultimate ambition. Over 95% of the

> *Earning a degree is not your ultimate goal; it is merely a means to an end, your rewarding career as a conservation professional.*

Saving the Earth as a Career: Advice on Becoming a Conservation Professional, Second Edition.
Malcolm L. Hunter, Jr., David B. Lindenmayer, and Aram J. K. Calhoun.
© 2016 John Wiley & Sons, Ltd. Published 2016 by John Wiley & Sons, Ltd.

time pursuing these two goals – degree and career – will be entirely consistent, but some exceptions may occur. For example, should you dedicate the whole weekend to gathering more data for your thesis or should you spend Saturday helping some local conservation professionals on one of their projects? The latter may have virtually nothing to do with earning your degree, but might be a worthwhile step on the road to your career, be this for résumé-building, networking, or simply expanding your educational horizons. Balancing all the demands on your time is never easy – as we will see later in this chapter – but it will help to keep your eye squarely on your goal, and to be sure that your advisors are aware of your goal.

If they are not reminded of your career goals, occasionally some faculty members will drift into the assumption that you aspire to become an academic like them, especially if you are a doctoral student. Perhaps you do wish to pursue an academic career – certainly many academics make huge contributions to conservation – but there are far more jobs for conservation professionals in other sectors, and you should consider a wide set of options, as outlined in Chapter 9.

We will return to the subject of communicating with your advisor and advisory committee below, after reviewing the diverse elements that may coalesce to constitute your experience as a graduate student.

A PROJECT

The single biggest difference between being an undergraduate and a graduate student is that you will probably need to undertake some kind of major project, in many cases research leading to a thesis. It is difficult to generalize about this because expectations vary enormously among countries, among universities within the same country, and among disciplines. There are some graduate programs that have no project requirement at all or demand only a modest literature review or practical experience working with a relevant organization akin to a conservation apprenticeship; that is, undertakings that can be completed in a few weeks or months. On the other end of the spectrum are full-scale research projects that require years of dedicated work from you (and perhaps a team of research assistants) and will consume over 90% of your overall effort.

Undertaking a large-scale research project is clearly an essential experience for someone who wishes to pursue a career as a conservation researcher, probably in an academic or government position. However, the skills required for completing a sizable project are also relevant in almost any professional post: developing and honing a project idea until it can be cast as a coherent proposal; executing a project that may involve complex logistics, staying within a budget, and collaborating with colleagues; analyzing data in a statistically and logically sound manner; and presenting your work in both verbal and written forms accessible to both professional and lay audiences. After surviving all this, any assignment that comes your way as a conservation professional will seem far more manageable. As you can surmise, we are great fans of undertaking major projects and the next three chapters are dedicated to designing, executing, and presenting outcomes from a project. Most projects will be some form of research, but much of the guidance we offer can be extrapolated to almost any independent body of work.

COURSE WORK

Unless you are in one of those all-research programs, you will continue to take course work as a graduate student. Of course, choosing the right set of courses is a unique process whereby you overlay the courses you have taken as an undergraduate, your particular career interests, the requirements of a specific graduate degree program, skills required for your project, and the courses available at your university or elsewhere (e.g., on-line courses and short courses). The set of courses you would ideally like to take will probably have to be trimmed because of the demands of your project. On the other hand, you will probably feel compelled to take some courses that you will not relish taking, but which are essential to your work (statistics often meets this description). Hopefully you can manage to take the essential courses plus some that appeal to you because of their excellent reputation and because they will diversify your knowledge. In particular, the multidisciplinary or interdisciplinary nature of conservation demands that you reach beyond traditional disciplinary lines such as biology or economics. At a minimum, be careful to meet all the requirements for your degree, especially as these change occasionally and your advisor may not be on top of the latest expectations.

A BALANCING ACT

If you are in a program that requires both a project and course work, then you have to think carefully about how much weight to give each, plus other demands on your time that we will discuss below. Be sure to discuss the issue with your advisor and other people before registering for your first set of courses, and it may be worth revisiting the subject briefly each term.

You need to develop a wide repertoire of professional skills and that often means focusing energies on a project rather than courses.

Most people strongly emphasize course work early in their program to clear the deck for project work later on.

With good planning you can put the balancing-act issue to bed, but it may wake you up again, crying for attention in the middle of the night, when you encounter a major time crunch. Imagine that it is 3 o'clock on Saturday morning and you are wide awake, unable to sleep because of your dilemma: you really need to spend the weekend looking for one more study site before your project begins on Tuesday, but you also have to study for a huge exam in Environmental Policy on Monday. What do you do? Hope you get lucky and find a great study site on Saturday? Dedicate the daylight hours to project work and consume lots of caffeine to study at night? There is no perfect answer, of course. Most graduate students have spent many years earning excellent grades and their inherent tendency is to stick with a winning formula: study hard for top grades. Standing at a distance it is not so obvious that the risk of only earning an 88 instead of a 95 on the Environmental Policy exam should outweigh the risk of not finding another study site. What if your study goes forward with only five replicate sites instead of six and that difference torpedoes the statistical significance of a key test of your main hypothesis? Ouch! Of course we are spinning tales to make a point here. Doing well in your course work *is* important; for example, your grades might be evaluated in the competition for a university-wide fellowship. However, course-work performance may not be as important as you think it is, especially if you weigh it against the success of a project.

> I don't usually discuss course grades with my advisees, but one day I asked one of my students why he seemed a bit down and he replied that he was disappointed with a test grade he had received that morning. I told him that it was not critical to get As all the time; that most employers don't even consider the grades of someone who has earned a PhD. "Oh I got an A," he replied, "I was just disappointed not to get the highest grade in the class." That led me into a short emphatic speech about looking forward to his career and not looking backwards at the kinds of milestones of success that had worked for him in the past.

Although pleasing your advisor is less important than furthering your personal development, it should be noted that your advisor will probably care much more about your progress on a project than your course grades. This may come in part because advisors have the perspicacity to put courses in context and in part because they feel an obligation to oversee a successful project, especially if it has been funded by an external agency that is eager to see concrete results.

Whereas course work and project work may be the two elephants in your kitchen, they are not alone. In the following sections we will consider some other activities that your degree may require, as well as other ways, unrelated to earning a degree, to foster your professional development. These activities may be overshadowed by independent project and course work but they can lead to some critical professional skills.

TEACHING

Some of your most memorable experiences as a student may come with a role reversal that puts you in front of a class as a teacher. This is because

Teaching can be a golden opportunity to develop your communication skills.

many graduate students earn their entire stipend and tuition from teaching assistantships, and many more try it for a term or two. Dozens of books have been written about how

to be a good teacher or mentor and your university may offer a short course to prepare its teaching assistants.[5.2] Here we just offer two basic ideas for using teaching opportunities to further your development as a conservation professional, based on the lyrics of an old song.

1 **Accentuate the positive.** No matter what your career trajectory, the communication skills you will learn as a teacher are invaluable. If you are speaking to an entire class you will gain poise as a public speaker. If you only interact with students one on one you will develop a greater understanding of how diverse people think and act, and hopefully become more empathetic and a better mentor. If you spend most of your time correcting papers, you can learn how to improve your own writing and how to provide constructive, positive criticism. Furthermore, being compelled to teach someone about primary succession or endangered species policy is an excellent incentive to learning this material very comprehensively yourself. In other words, you have to thoroughly understand something before you can explain it. Finally, if you have any authority for managing a class, then you will gain leadership skills and have the opportunity to inspire the next cohort of potential conservation professionals. So the bottom line is, recognize the ways in which teaching is an opportunity and exploit this opportunity as fully as possible. Of course, if you aspire to be a professional educator all this will be doubly valuable experience and good fodder for your résumé.

2 **Eliminate the negative.** First, make sure you discuss with the instructor what is expected of you as a teaching assistant before the course begins so

that no one is surprised mid-term. Perhaps the worst part of being a teaching assistant is correcting a huge pile of exams or reorganizing a laboratory recently vacated by a sloppy mob. Some of this may be unavoidable, but you may be able to shift your burden toward more rewarding work if you volunteer for it. For example, you could offer to help design the exams not just grade them. Better yet, volunteer to give one or two regular lectures or perhaps present a 10-minute case study illustrating the key points of a lecture. Do a great job with these and more opportunities for additional time in front of the class will probably follow. If you are brave, you can volunteer to teach an entire independent course, perhaps one of limited scope like a short field course or a readings seminar that meets once per week, or take one on as an equal partner with a faculty member. You might even have the opportunity to assume responsibility for a full regular course, perhaps substituting for a faculty member who is away on sabbatical. The amount of work involved in teaching a course for the first time is not trivial; indeed, substituting for a vacant faculty member could set your program back by months. Nevertheless, if it meshes well with other aspects of your program it is likely to be worth it, especially if you wish to pursue education as a career.

INTERNSHIPS

Some degree programs, primarily those that emphasize course work rather than project work, require their students to experience the life of a conservation professional by working for a conservation organization for the summer or during a term or two. These are often a not-for-profit conservation group or a government agency, or sometimes a for-profit group such as a consulting firm. The basic principles described above under Teaching apply here: focus on developing useful skills and volunteer for tasks that will help you gain these. For tips on landing an internship with lots of potential for learning see Chapter 9, Finding a job.

It may be tempting to take an internship lightly – especially if it is only a couple months out of a 2-year program – but that would be short-sighted. Many organizations use their internship program as a recruiting tool, so performing well could be a critical step toward long-term employment. On the other side of the coin, it could be a great opportunity for you to decide if working for the city's Planning Department or the province's Department of Natural Resources really is your dream job. More immediately, your work as an intern might lead to the development of a project that would be supported by the organization and have a significant impact on an important conservation issue. Finally, an internship can give you a sense of pride that you are advancing the cause of conservation as one small cog in a much larger machine.

COMPREHENSIVE EXAM

So you thought the final exam for genetics was a challenge ... well, some degree programs have a process for evaluating your knowledge of your entire discipline: for example "1. Describe the evolution of theoretical ecology in the 20th century and how it shaped conservation efforts. Use specific examples from at least three continents. You have 30 minutes; use at least 30 citations. 2. Rewrite your answer to question 1 by envisioning what global conservation would be like without the rise and partial fall of communism during the 20th century; 10 minutes, no citations necessary."

Often this is called a comprehensive exam because it covers such breadth; sometimes it is called a qualifying exam because it happens early in your program and is used to decide whether you are qualified to continue. These exams may be oral or written and are often both; written exams may be open-book, or closed, or both. Each program has its own exam culture so it is best to discuss it with your advisor, your examiners, and fellow students, especially those who have been through the process. You might even be able to witness an oral exam, although they are usually closed. In addition to format and content details, it is useful to know how much time most students spend in preparation, the portion of students who pass on their first attempt, and the process for retaking the exam. If they are available, reviewing copies of past exams is also useful. If you are having difficulty understanding some material, do not be shy about seeking assistance.

Poor performance on comprehensive exams is often a matter of being too nervous rather than ill-prepared, but it is rather unhelpful to offer the advice "Don't panic!" Relaxing the evening before the exam rather than further cramming might help. It is also useful to realize that your examining committee genuinely wants you to do well. In oral exams they may ask you many questions that you cannot answer, but that is almost inevitable. They are trying to map the boundaries of your knowledge and in trying to follow this line they are bound to ask many questions that step outside it. If they only asked easy questions they would not know if they were a centimeter or a meter inside the borders of your knowledge. They are also interested in intellectual synthesis; that is, your ability to transcend memorizing basic concepts to generate original ideas. This careful assessment of your knowledge may allow your examiners to suggest topics for which you should do additional reading. Finally, although comprehensive exams may seem nothing but torturous, try to think of them as a milestone in your intellectual development. Indeed, many students have finished the process and felt thoroughly immersed and current in their discipline, or at the "top of their game."

> *One student brought a blueberry pie to his comprehensive exam for the committee to enjoy. When asked about this afterwards he confessed that it was part of his strategy. He figured the examiners would not look so intimidating if their teeth were stained blue.*

INVESTING IN YOUR DEPARTMENT AND YOURSELF

As a new student it is probably premature for you to be trying to revolutionize the life of students in your program, but it is not too soon to start thinking about the issue. Ultimately, you may be able to improve the learning environment and foster camaraderie for everyone and, just as importantly, cultivate your own development.

Here are some questions to consider as a starting point.

- Does your department have a regular seminar series with outside speakers? If so, do the students have an opportunity to select and invite speakers, or meet with them without the distraction of a faculty member? Does your seminar series ever include interdisciplinary seminars with other departments?

- Are there informal but regularly scheduled gatherings to talk about relevant subjects; for example, recent papers of interest, current conservation issues, ethical or philosophical topics?

- Are there organized sessions where you can discuss life after graduation? In particular, what are the expectations of employers, and how you can prepare yourself for them? In the same vein, are there opportunities to meet representatives from a variety of organizations to discuss careers in conservation?

- Does your department have a set of written guidelines to help you find your way through the university bureaucracy, or take advantage of university resources such as a job center or writing clinic?

- Are students represented at departmental faculty meetings or on university committees? Do they have a role in hiring new faculty?

The difference between graduate student roles in departmental governance is night and day between this university and the one where I did my doctorate work. Overseas we were barely aware that the department was hiring a new faculty member. Here a graduate student representative has a formal vote on the selection committee, the students spend hours with each candidate, and their collective ranking of the candidates has enormous weight in the ultimate decision.

After you understand the lay of the land, you can start discussing these questions and others with your fellow students and any faculty members who seem particularly empathetic. If there is broad support for a new initiative, then a small group of students could meet with the department chair. The next step might require a brief proposal to share with the entire faculty.

Of course, some initiatives do not require all these steps. For example, if you want to invite a group of people to have lunch together on Wednesdays to discuss recent journal papers or interdisciplinary issues in conservation, you do not need formal approval to do so. Discussion groups are quite easy to organize and often rewarding. One difficulty is that they can be stifled by their own success if so many people attend that this constrains people, especially shy members, from speaking. In a large program it might be necessary to structure attendance by having multiple discussion groups and a sign-up sheet to distribute people among them.

Some of our students organized a very successful series of mentoring sessions, each week inviting a different faculty member out for a meal during which they would informally discuss topics such as professional ethics and various key skills needed for a successful career. Three of the students became so interested in the issues that they published a paper on the topic.[5.3]

Obviously some of these ideas – such as mentoring sessions with conservation professionals – can represent a significant boost to your professional development, something quite distinct from course and project work. Furthermore, working with other students to improve the overall educational environment may help you develop a skill that might otherwise be neglected in your personal development: working with a team. Perhaps you have a long history of being active in team sports or working on group endeavors, but many people who gravitate toward the natural world have a tendency to be lone wolves. At some level, you are who you are, and you cannot recast your personality. However, your efficacy as a conservation professional will certainly be enhanced if you learn to work with a group to pursue a common goal and collaborating with your fellow students, or even leading them, is one good opportunity.

We must end this section with one caveat. Some students find participating in departmental activities so rewarding that they compromise their core work. Some activities will engage you every week, like attending a departmental seminar series; others you should try just once, such as serving on a faculty selection committee; in total all of these activities should consume only a modest portion of your time, the icing on the cake of your graduate experience.

EXTRACURRICULAR ACTIVITIES

So, are you beginning to feel like a juggler with all these balls that you have to keep aloft? Juggling can be quite stressful, especially if dropping a ball has serious repercussions, and thus you need to take some breaks. That is not always easy to do. All three of us have experienced limited periods, especially during field seasons, when work demanded incredibly long hours for day after day without a break, but that is emphatically not how we live our lives on a regular basis. You have heard all the exhortations about a good diet, sufficient sleep, and regular exercise and we need not repeat them here except to say that it's true and your success as a conservation professional – and enjoyment of life – will be compromised if you do not maintain your physical and mental health.

We can offer a few thoughts, in no particular order, about how to combine your need for diversions with your career aspirations.

- Enjoy the outdoors. Exploring natural ecosystems is a superb way to relax, get exercise, and to become a better naturalist. Undertaken with a few friends it is also a wonderful way to socialize. Outdoor-oriented clubs, either on campus or in the local community, can help you explore the nearby region. Some of these organizations may have a conservation agenda; we will return to this issue in Chapter 10. Spending time in the outdoors is often also a way to reinforce the reasons why you have taken a career path in conservation.

- Help other people with their field work. It is a great opportunity to learn about their work, socialize, and get some exercise.

- Socialize widely. If you are in an environmental economics program, reach beyond the other economists and get to know some marine biologists or soil and water scientists. You may also learn some interesting and useful things from historians and artists. Similarly, conservation is very much an international endeavor and universities are often wonderful places to meet people from other nations and obtain a small taste of how they perceive the world. If you're not sure what to say to new people, ask them a question about themselves or offer a joke. How about this one:

> *A sociologist, a biologist, and a statistician were traveling by train through the countryside admiring herds of black cows. As they crossed a river into a new county they spotted a white cow and the sociologist remarked, "Isn't that interesting; the cows in this county are white." The biologist responded, "My goodness, let's not jump to hasty conclusions. All we really know is that some of the cows in this county are white." To this the statistician replied, "Hasty conclusions? All we really know is that one half of one cow in this county is white."*

- Walk, run, or cycle to and from work because it is good exercise, can be a chance to either relax your brain or focus it on an interesting idea, and does not generate air pollution.

- Read fun books about the natural world. Most of the books you will read as a student will do little for your mental health beyond helping you to fall asleep. Under Further readings[5.4] we have listed some of our favorite books about nature and conservation, selected with the key criterion that they be enjoyable to read; there are hundreds more of course. Share the books with friends and colleagues as the basis for interesting discussions.

- Keep a journal or blog. Record your thoughts about your work experiences. Writing helps you reflect on your experiences and often reveals new ideas. It even keeps you limber for writing your thesis. Some people augment their journals with photographs, mementos, or artwork.[5.5]

We do not mean to imply that a conservation professional shouldn't relax by reading a Stephen King novel or exercise by dancing to hip hop. Diversity is the spice of life! These are just a few ideas of how you might have fun and also advance your career.

COMMUNICATING WITH YOUR ADVISOR

Advisor–advisee relationships vary dramatically. The best ones grow through mutual respect and admiration into a life-long bond of being colleagues and sometimes friends. The worst ones are too horrific to contemplate. Because every person is unique, the intersection of any two people defies

Take a proactive role in establishing an efficient means of communicating with your advisor.

attempts at generalization. Of course, the same is true of marriage and parenthood, and hundreds of books have been written full of generalized advice about how to make those relationships work. We will cover the advisor–advisee relationship quite succinctly here, in part because it is best discussed in subsequent chapters, in part because most of its success is based on principles that are true of all human interactions. Honesty, openness, respect, empathy, loyalty, kindness, patience, and so on are just as important to a strong relationship with your advisor as with your friends, spouse, parents, siblings, or children and thus need little elaboration here.

We suggest that you take a proactive role in developing the relationship. Just because your advisor may have more seniority and authority does not reduce you to a frail violet waiting for attention. Act like a junior colleague rather than a timid student and you are more likely to be treated like a colleague. Of course, the main thing you need from your advisor is advice and this will probably come in three basic forms: verbal during meetings and phone calls, written as feedback on documents you write, and e-mails, those odd hybrid conversations/letters. Observe your advisor's work schedule and this will help you gauge the best times for setting up meetings or requesting a manuscript review. One of the key things to ascertain is the logistics of communicating with your advisor.

Pat: "Thanks for all your thoughts Dr. Black. Just one more question for today: should we have regular meetings?"

Dr. Black: "First, please just call me Terry. We can meet whenever you want to set up a meeting, or Tuesday from 12 to 12:30 should fit both our schedules. Thirty minutes may not seem like much but if you come with a specific set of issues you'll be surprised how much ground we can cover."

Pat: "And what if I have a quick issue that can't wait?"

Terry: "E-mailing is fine, or any time my door is open just pop in quickly"

Pat: "What about documents like my proposal? Do you want those e-mailed or in hardcopy?"

Terry: "I'll probably be pretty intense on the first draft or two and that's easier in hard copy. We can switch to e-mail when we are polishing later drafts."

Pat: "Any idea how long it will take you to do a review?"

Terry: "Of course, that depends on lots of things, but if you give me some forewarning I might have it back in a couple days. If it arrives out of the blue during a busy period it could be easily take over a week."

Pat: "Okay, that covers my issues for now but I'm likely to have others come up over the next few days."

Terry: "That's fine. Happy to answer them if I'm free and you'll find the other students in your office know the ropes very well too. Now let's go down to the department office so I can introduce you. Having a good relationship with those folks is very important. They're always better at answering administrative questions than I am."

There are other many other aspects of working with your advisor that will arise in the next few chapters, but none is more important than establishing a system of communication that is efficient for both of you, so do not be shy about pursuing this early in your program. A rusty memory is one of the worst impediments to efficiency, so it is often wise to have a written list of issues before meeting with your advisor and to take notes on the resolution during or immediately after the meeting.

Occasionally students end up with two co-advisors who share the role equally, perhaps because a research project demands their complementary skills, perhaps because they bring complementary resources to a project such as an assistantship and an appropriate laboratory. This arrangement can work well for all three parties, but it does require even more attention to how communications will work. For example, it may shift the weight from spontaneous face-to-face communication to scheduled meetings and e-mails in which three people can easily participate.

AN ADVISORY COMMITTEE

Your advisor might be a very bright star in your academic cosmos, but there are other stars too. In many programs you will work with your advisor to identify two to five other people to be your official advisory committee or panel. They will probably have some role in shaping all the decisions discussed above, but typically they are primarily chosen because their expertise is relevant to your project, perhaps complementing your advisor's talents.

Do not take these selections lightly. A poor committee member is useless at best and at worst a serious impediment to progress. Ask other students for opinions on effective committee members; specifically, you want people who are insightful critics without being nasty, available for consultation, and responsible about reviewing proposals and papers in a timely manner. You do not need to select the friendliest faculty members, although your decision may be influenced by personality issues. In particular, you do not want to have a committee member who dislikes your advisor. Similarly, while good-spirited disagreement can be wonderful fuel for creativity, you do not want a committee member who has little respect for your advisor.

Logistically it will definitely be easiest if all your committee members are in the vicinity – around campus or at a nearby institution. That said, electronic transmission of documents, speaker phones, video conferencing, and other technologies might allow you to find a committee member anywhere in the world. If you go outside the university, you may need to help organize adjunct faculty status for this member, but this is usually not too complicated. If you want to strengthen the conservation aspect of your work, and perhaps facilitate getting a job when you graduate, you should consider inviting a conservation practitioner to be on your committee. Even if they do not have the credentials that the university requires for an adjunct faculty member, they could still serve on the committee in an *ex officio* status. If your university does not require you to have an advisory committee, you might still consider forming an unofficial one, even if it only meets once to discuss your research proposal. Ideally a committee will continue to meet, once or twice a year is common, to review your progress and discuss mid-course corrections. On the other hand, it is not necessarily essential that your committee ever meet as a group. Although far less than ideal, you could solicit their input through a series of one-on-one meetings and keep them abreast of your progress with occasional written reports. Whether it is a quick consultation with your advisor or formal meetings with the whole committee it is usually best if you are chief organizer; no one is as focused on your progress as you are.

Whether it is a quick consultation with your advisor or formal meetings with the whole committee it is usually best if you are chief organizer, because no one is as focused on your progress as you are.

WHEN THINGS GO VERY WRONG

In the famous words of Robert Burns, "The best-laid schemes o' mice an' men gang aft agley" or in more contemporary, pedestrian language: "Good plans are often trashed." Almost every student experience will include some challenges, like a rejected paper or a personal crisis, and we will discuss coping with some of the more common problems in the chapters to come. In this section we want to cover the larger issue of what to do when things are going so badly that you seriously consider quitting: for example, you barely pass your first set of courses, your advisor resigns and leaves academia, you become seriously ill, or your partner moves to another city.

The first thing to do is talk about the problem with some other people: friends, family, your advisor, other students and faculty members, and professional counselors. Many universities offer free mental health counseling with people trained to deal with many problems. The right mix and sequence of

people will vary depending on the issue and the nature of your social network (obviously discovering a fatal flaw in your project is profoundly different than being the victim of sexual harassment). You are not the first student to contemplate quitting and some of these people, especially your advisor, may be able to suggest some options that will surprise you. Perhaps you can take a leave of absence and return later; perhaps you can switch to a less demanding degree; perhaps you can change projects or even advisors. Your advisor and university have made an investment in you that they will not want to lose; they do want to see you succeed. Once you understand your options, take the time to weigh them carefully. Listen to what other people have to say, but in the end it is your life and your decision.

Pat:	"I have some big news to share; Chris and I are going to have a baby in May."
Terry:	"Wow; that *is* a surprise."
Pat:	"Yeah, it was a surprise for us too. We know that the timing is terrible with field work beginning in May too; I don't see how it can possibly work."
Terry:	"No, I don't either, given the travel difficulties."
Pat:	"That's why I think I'll have to quit the program."
Terry:	"Well don't leap to that conclusion yet. There may be some other options."
Pat:	"The problem is not just having the baby. The only way this can work financially is if Chris continues full-time work and I stay home with the baby."
Terry:	"Hmmm, that certainly limits your options. I suppose you could always switch to a non-thesis degree. Don't you already have almost all the course work you would need?"
Pat:	"Yes, I could probably swing that. My parents suggested another option. If I could transfer to Lambda University and move into the apartment they have over their garage, then they would be willing to help out with child care."
Terry:	"That's definitely worth looking into if Lambda will take transfer credits."
Pat:	"Okay, let me discuss all this with Chris over the weekend and get back to you on Monday."

Terry: "Oh by the way: Congratulations! We dug into finding a solution so quickly that I forgot to congratulate you."

Pat: "Thanks. I do appreciate you being so understanding. I've never quit anything important in my life so I am feeling awfully guilty."

Terry: "Well, to be honest I'm not happy at the prospect of losing you and having to find someone else, but things like this happen, and it doesn't make you a failure. At some point you'll get back on track with your career and hopefully your time here will be useful to you."

One situation merits special mention here: what do you do if you want to change advisors? Sometimes it becomes clear to both advisor and advisee that a change would be best and you should be able to negotiate this directly and amicably. However, if your relationship has gone sour you may need to discuss your options first with someone like the chair of the department or the coordinator of graduate students. You will definitely not be the first person to come to them with this issue. Leaving your advisor might mean leaving your degree program and the university, but there is a good chance that alternative arrangements can be made for you to finish under another advisor in the same department or in a different department at the same university. Because the university has made an investment in you there will be some level of interest in helping you through to completion and you need to assess that path.

In summary, whereas there is a bit of insight in sayings like "When the going gets tough, the tough get going" or "Where there's a will, there's a way," every situation is unique and we can only suggest that you thoroughly discuss options with other people. Discuss options with other people but ultimately the decision to continue, radically modify, or discontinue a degree program is your call. The sooner you make a decision the better, but do not act rashly. If you decide to depart, you will be following a trail left by many other students, some of whom shifted to very different careers, some of whom found another route to being a conservation professional.

> *Discuss options with other people but ultimately the decision to continue, radically modify, or discontinue a graduate program is yours.*

A FINAL WORD ON WORK STYLES

Some successful people make their mark in the world mainly by working very hard, some by being very talented. Most successful conservation professionals combine a solid capacity for work, various forms of natural intelligence, and a passion for conservation. If you really enjoy long hours of work then that is a straightforward path to success. If, like most people, your capacity for work has some bounds, then you need to think strategically about your work, figuring out ways to be efficient. Striving to think as hard as you work, or to work smart, is a recipe for both increased productivity and greater enjoyment of life. Universities, with their multiple, bottomless opportunities to allocate your time, are a very good place to learn to be efficient. Ultimately, these time-management skills extend far beyond your university years – in essence, they are an investment in your future, both as a successful conservation professional and in a lifelong quest for personal fulfillment.

Striving to think as hard as you work, or to work smart, is a recipe for both increased productivity and greater enjoyment of life.

Further readings

5.1 A number of articles give students advice on designing their program, often covering topics that are widely distributed in this book. Here are some citations where you can read other people's perspectives.

Blickley, J.L, K. Deiner, K. Garbach, I. Lacher, M.H. Leek, L.M. Porensky, M.L. Wilkerson, E.M. Winford, and M.W. Schwartz. 2013. Graduate student's guide to necessary skills for nonacademic conservation careers. *Conservation Biology* **27**, 24–34.

Campbell, S.P., A.K. Fuller, and D.A.G. Patrick. 2005. Looking beyond research in doctoral education. *Frontiers in Ecology and Environment* **3**, 153–60.

Colon-Rivera, R.J., K. Marshall, F.J. Soto-Santiago, and D. Ortiz-Torres. 2013. Moving forward: Fostering the next generation of Earth stewards in the STEM disciplines. *Frontiers in Ecology and the Environment* **11**, 383–91.

Perez, H.E. 2005. What students can do to improve graduate education in conservation biology. *Conservation Biology* **19**, 2033–5.

Schmidt, A.H., A.S.T. Robbins, J.K. Combs, A. Freeburg, R.G. Jesperson, H.S. Rogers, K.S. Sheldon, and E. Wheat. 2012. A new model for training graduate students to conduct interdisplinary, interorganizational, and international research. *Bioscience* **62**, 296–304.

5.2 Two books we recommend are

Bain, K. 2012. *What the Best College Students Do*. Harvard University Press, Cambridge, MA.

Bain, K. 2004. *What the Best College Teachers Do*. Harvard University Press, Cambridge, MA.

5.3 See the article by Blickley et al. 2013 under note 5.1.

5.4 A sample of some good books about the natural world: *At Play in the Fields of the Lord*, Peter Matthiessen; *Song of the Dodo*, David Quammen; *Song for the Blue Ocean*, Carl Safina; *Pilgrim at Tinker Creek*, Annie Dillard; *The Last Panda*, George Schaller; *Lost World of the Kalahari*, Laurens van der Post; *Tracks*, Robyn Davidson; *Gift from the Sea*, Anne Lindbergh; *Shadow of the Condor*, David Wilcove; *Throwim Way Leg*, Tim Flannery; *Guns, Germs, and Steel*, Jared Diamond; *Last Chance to See*, Douglas Adams; *A Year in the Maine Woods*, Berndt Heinrich; *Living on the Wind*, Scott Weidensaul; *Diversity of Life*, E.O. Wilson; *The Beak of the Finch*: Jonathan Weiner. We have only listed one book per author, but most have produced multiple books that you are likely to enjoy.

5.5 For an example of a journal page, see

Jacobson, S., M.D. McDuff, and M.C. Monroe. 2006. *Conservation Education and Outreach Techniques*. Oxford University Press, New York, NY.

6 Designing and executing a project

If you are in a degree program that requires a research thesis or another signif-icant project, then this chapter is where your journey along the trail reaches the base of the mountain and the climbing begins. The ascent will be much easier if you choose the right route. Thus, the first half of this chapter is devoted to the all-important issue of planning your project. Much of this chapter leans toward undertaking a thesis research project but, in broad brush strokes, the core ideas are highly relevant to developing any significant project, for example develop-ing a management plan for a tract of land, or a community-based conservation program.

SELECTING A TOPIC

Many factors will influence the topic you pursue for your project: your interests, the expertise and interests of your advisor or employer, the value of the work for conservation, the funding available to support the work, or the type of graduate scholarship or fellowship that is being offered to you. For some graduate-student positions, funding is strictly allied to a grant for which the topic is "set in concrete" (Chapter 3). In other circumstances, there will be more flexibility in selecting a topic.

You should consider what skills you will develop by pursuing a particular topic. Many project-based skills are generic career-long skills – for example, how to conduct research, how to manage a large project, and how to write a professional paper – and can be applied widely. Others skills are much

Saving the Earth as a Career: Advice on Becoming a Conservation Professional, Second Edition.
Malcolm L. Hunter, Jr., David B. Lindenmayer, and Aram J. K. Calhoun.
© 2016 John Wiley & Sons, Ltd. Published 2016 by John Wiley & Sons, Ltd.

more specific. For example, if your thesis research is centered on the use of microsatellite markers to address conservation genetics, it is unlikely that these skills would be particularly useful if your next job requires using contingent valuation to quantify the economic benefits of conserving endangered species.

SETTING REALISTIC EXPECTATIONS

Terry: "Well, you have had some time to think about how your thesis work will fit with our laboratory. What things really interest you?"

Pat: "I'd be really keen to study Patagonian Tigers – they are such stunning animals. I'd like to radio-track them to compare movements and social behavior among seasons; that's never been done before. I could also collect scats for diet analysis and correlate that with surveys of prey populations."

Terry: "Hmmm, certainly conservation work on Patagonian Tigers would be valuable and would fit with our lab. But let's chat about what can realistically be done over the next 3–4 years."

Pat: "Just looking at your face, I thought you might say that."

Terry: "It's true that radio-tracking can be useful, and I hate to be a killjoy, but there are practical issues. What do the radios cost now?"

Pat: "To cover everything would be about $US950 each."

Terry: "So to track 50 animals you will need almost $US50,000."

Pat: "I know that's a lot of money, but aren't the large cat conservation groups loaded with research dollars?"

Terry: "We rarely have got more than $20K for one project from them and you would need a lot more than that just for logistical support – one or two assistants, maintaining a field camp, transportation – before you even bought your first radio-collar. There are also some ethics issues here. You are going to need to sedate animals and for that you will need a wildlife vet. Otherwise, rare carnivores dropping dead on you would be a public relations disaster."

Pat: "Perhaps we need a rethink on this."

Terry: "I agree."

This discussion embodies some key issues that characterize many discussions between advisors and students when a new project is being initiated. New students often need to have "their wings clipped" somewhat to ensure that a good project can be realistically completed with the time and resources available. As you design your project, keep three points in mind.

1 **You only have to finish one degree.** Almost every student starts on a project with an unrealistic expectation of what they can accomplish. Many propose work programs that could easy fill two or three projects. Similarly, most students fail to realize that even if you have 2–4 years for a project, tasks usually take longer than expected, in part because unpredictable problems arise.

2 **It is better to do a few things well, rather than a lot of things poorly.** A good project is often one that contributes new and deep understanding to a topic.

It is better to do a few things well, rather than a lot of things poorly.

It is rarely possible to do this if you are spread too thinly over a broad issue. As with many clichés, there is much truth in the one "Nothing that's worth doing comes easily."

One of the most difficult theses I have examined was one that had 14 chapters, with 12 of them briefly touching on a wide range of issues. None of them scratched much below the surface and I felt compelled to return the work and recommend a major additional period of more detailed work focusing on a subset of questions. I blame the advisor for this as the thesis seemed to show that the student was confused about what a good research program might entail.

3 **You need to be passionate about your project.** The up-side of the conversation between Terry and Pat is that Pat is clearly very passionate about a Patagonian Tiger project. This is important and Terry needs to work with Pat to direct that passion into a creative and stimulating project that is achievable and useful. There will be many times during your program that you will feel disillusioned and question the value of what you are doing. This is entirely normal. But it is important to be very interested in the work you are doing to help overcome the low points. Students who are simply doing a project for the sake of getting an advanced degree are often those that struggle the most when things are not going to plan. For a person who is passionate about a particular topic, enthusiasm and dedication will ensure that "the work won't seem like work."

Over the years I have had students who have triumphed over adversity – divorces, illness, and other traumatic events. In each case, the student completed their project eventually in large part owing to their passion for the work.

FRAMING THE PROBLEM

A problem well stated is a problem half solved. (C.F. Kettering, 1876–1958)

Once you have selected the broad topic on which you will work, it is essential to define the problem you are going to address more specifically. Ecologists

and conservation scientists are notoriously bad at doing this.[6.1] Perhaps it is because ecological systems and, by extension, conservation programs, are so complex that they generate a "blizzard of ecological details." Whatever the reasons, carefully defining the problem is fundamentally important and often this will require addressing questions such as these:

• Will answers to these questions advance science and conservation?

Ideally one would like both to push intellectual frontiers and to make a solid contribution to conservation. Unfortunately, you will probably have to favor one over the other. For example, if you are studying the nutritional physiology of a rare bird species, you might make an important contribution to animal physiology, but your work will probably have little

Ideally one would like both to push the intellectual frontier of science and to make a solid contribution to conservation.

conservation impact if the bird's populations are declining because of human exploitation. Conversely, studying how management by local stakeholders can mitigate overexploitation of this species may only add one small brick to the foundation of science, rather than take it in exciting new directions.

This latter idea is captured in the idea of "aardvarks and Arcadia".[6.2] People who aspire to cutting-edge research try to avoid being guilty of the "aardvark principle" whereby research is viewed as novel just because it is undertaken on a new species. For example, if lion, tiger, and bear populations have been shown to exhibit functional responses to food abundance, it is not very novel to discover that aardvarks do too. Conservation scientists should not lose too much sleep over this issue because every species is unique at some level, and thus if it is important for the conservation of aardvarks to understand how they respond to changing food abundance then that is adequate justification. The issue is a bit fuzzier when we turn to the Arcadia principle, whereby researchers would argue that they are doing ground-breaking research because they are studying aardvark responses to changing food abundance in Arcadia, whereas previous research on the topic was undertaken in Xanadu and Shangri la.

At some level of detail, ecology, socio-economic systems, and conservation are unique to a particular place and this can justify undertaking the same type of project in multiple locations. However, this is often done thoughtlessly with much waste of resources. For example, if the conservation policy makers of Arcadia won't accept research done in Xanadu even though the border separating the two is a human construct, then time and dollars have been squandered. If you are going to repeat an old study in a new place be sure that the rationale for it is sound, and use methods that will allow you to directly compare your work with previous efforts.

Your advisory committee will be able to offer solid advice on whether or not your research will advance fundamental knowledge. They may or may not be in a good position to determine if your research will advance conservation. For this perspective, it would be wise to talk to some conservation practitioners who deal with the issue you will address on a regular basis.

• What other work has been done on my topic? How can this help guide me?

This is a critical step. A key part of designing any project is reading the background literature to determine who has done what, where, and how. There is a lot of reinventing of wheels because people do not read and appreciate the past literature.[6.3] Don't worry that "it's all been done before" – that is never the case. There is always far more creative conservation work to do than there are people to do it. You just need to read the literature to find a suitable niche for your work. Moreover, it will be impossible to publish your work in journals if you don't cross-reference your work to previous efforts. Of course there is a veritable mountain of literature out there and it is growing at an increasingly rapid rate, so you won't be able to read it all before you actually begin. Ongoing reading will be essential throughout the duration of your program.

• What kind of information will I need and how will I obtain it?

This requires some deep thinking about information needs, particularly if you need to gather original data, as is the norm for thesis research. In this case, linking your methods, experimental design, and the key questions you are asking is a challenging, iterative process that takes a lot of energy to resolve. Some of these issues can involve as much statistics as conservation science.

This highlights the value of combining clear scientific thinking with appropriate statistical support, especially for experimental design. Qualified statisticians can help with planning experiments and observational studies, the kinds of statistical methods to be used in data analysis, and interpreting the results of data analyses. Indeed, statistics is a field evolving much faster than conservation science and is a highly credible discipline in its own right. Therefore, seeking expert assistance at the beginning of a project is much better than attempting to rescue a flawed design at the end. Remember that few statisticians will have any scientific experience and what they recommend, for example, as ideal sample sizes, may prove to be impossible to obtain in the real world. Thus, you may need give-and-take discussions about the realistic statistical solutions to the key conservation questions you have posed.

A new student arrived in my lab after working in a government agency where the culture demanded that you always be doing something tangible. She wanted to be out in the field within the first week of commencing her program. Thinking deeply about a set of questions and carefully framing a problem was not seen as real work in her former job. In the end, it took nearly four months for her to define her problems and execute what proved to be an excellent study on the conservation of farmland amphibians.

- How feasible is the project?

It is often valuable to carefully envisage the information that you will gather and assess whether it will be sufficient to actually address the key issues you are tackling. In the context of gathering original research data, sometimes a good way to do this is to construct a "dummy data set" (e.g., using a random number generator) to explore these ideas further. This approach might also force you to think about how much data will need to be collected to enable you to rigorously answer your questions. A pilot study can be useful to assess project feasibility. There can be nothing worse than spending months or even years in difficult field conditions to discover at the end that you simply don't have enough information to be confident that your results are meaningful.

I recall working with a student on a design for a major edge study. The ideas were great, the questions were fascinating and important, and the experimental design seemed sound. I suggested it was time to go the field and starting scoping out the project on the ground. Only a handful of the "right" kinds of sites actually existed; replication was impossible. Back to the "drawing board."

For relatively large projects, such as a multiyear doctoral research program, it can be useful to consider risk management. That is, the thesis might have some low-risk components for which you are confident that high-quality data can be gathered quickly. It also might also have some high-risk components where it is harder to collect data, but rewards might be greater if the work succeeds.

Many conservation projects involve fieldwork that is a hard slog. Therefore, sometimes it can be very valuable to do a small pilot study early on to determine if the grand ideas you had back in the office will actually work in the field – "virtual reality can quickly descend into real stupidity." For example, social scientists often initially work with small focus groups to help inform broader-scale surveys. Another key issue with designing fieldwork is the selection of study areas. It is important to weigh carefully the time and money required for access, permission to use sites and perhaps manipulate them, the number of replicate sites available, and the similarity of sites that are meant to be replicates.

Projects that have a technical side can take much longer to complete than initially anticipated: "if it can go wrong, then it often does." Animals prove hard to catch, a fire kills all your tagged trees, human subjects do not respond to surveys, or you cannot get your protocol approved by the university's ethics committee, and so on. You need to make allowances for things to go wrong. Of course, there is great value in communicating with other people who have worked in similar situations and to learn from their experiences. This interchange may be as little as a couple emails or as much as a visit to their lab or field site.

In some cases, you need to assess a method with respect to its return on time invested compared to an alternative approach. Radio-tracking is a classic example. It is a very widely used technique but can be very time consuming. It usually generates point location data recalculated as home range sizes or activity patterns that may not be suited for sophisticated statistics. Thus you should ask yourself whether other methods, for example track counts, might be a better option.

WRITING AND PRESENTING A PROPOSAL

You may be required to prepare a formal proposal for your project. Rather than seeing this as a great burden, embrace it as an opportunity to be sure you have a well-designed project. In fact, the thinking and planning that goes into crafting a proposal is so valuable that we recommend preparing one even if it is not a program requirement. Although you may be anxious to start your project, there are many advantages to preparing a proposal. In particular, writing a proposal forces you to morph general ideas about saving the Earth into reasoned objectives. It requires a level of detail you may not have given to project execution in the absence of having to document each step. In other words, proposals force you to frame the problem in a more formal fashion and allow you to assess whether or not your proposed work can realistically (financially, logistically, and temporally) be completed during your short tenure.

Having a proposal also allows you to solicit reviews from other students, committee members, or other willing reviewers in a much more substantial way than through casual conversations in the hallway or over tea. Feedback early in your program is critical and often helps you to eliminate problems you may not have anticipated.

Writing a proposal forces you to morph general ideas about saving the Earth into reasoned hypotheses and objectives.

Finally, because a proposal has many similarities to proposals for funding agencies, this can be a great opportunity to practice the important skill of writing a proposal to "sell" your ideas to other people. Proposals for funding must be well researched, compelling, and realistic, or they will be rejected. Imagine that you were sitting on a review panel with hundreds of proposals to consider, would you choose yours as one of the small minority to receive funding?

Elements of a proposal

Expectations for what a proposal should include may vary among departments or programs. Be sure to ask your advisor for both guidelines and samples of successful proposals from previous students. Our advice on what you might include in your proposal follows.

Introduction

> "I propose to study the effect of increased water temperature on reproduction of a sea star (*Asterias indica*) in the Indian Ocean. I am the first person to ever consider sea star reproduction and changes in water temperature due to global warming. Spiny and Echino (1986) have shown that sea star leg regeneration is dependent on water temperature in the Indian Ocean. Metoo et al. (1999) found the same thing in the Pacific. Water temperature may also affect exoskeleton hardness (Derm 1992) ... and finally, several species of sea stars may be imperiled in tropical waters."

This is not taken from an actual introduction. However, the problems illustrated here are common and to be avoided like the plague. The Introduction is your opportunity to capture the interest of the reader. You have to present a compelling question or problem, set in the context of current knowledge and a conservation problem.

Use the upside-down pyramid method. Start with the broader conservation problem and work your way down to what your project will do. For example, the Introduction would set the stage better if it began with two sentences such as: "Global warming is widely recognized to be a major threat to a wide variety of species and ecosystems" and "Among the species that may be most in jeopardy are several species of rare sea stars that inhabit tropical oceans." Scan the Introductions of some journal articles to see how they establish a broad and interesting context for the paper right at the beginning.

An Introduction should contain a concise, synthetic literature review of your study topic. It is not the place for a comprehensive literature review explaining the results of each paper ever written on the topic. For example, "Most of the research on water temperature and seas stars has focused on anatomy (Spiny and Echino 1986; Derm 1992; Metoo et al. 1999)" is more concise than a recitation of each paper. Only go into detail if it is necessary to show how your proposed research fits into the context of what is currently known.

Finally, the Introduction should end with an explicit statement of your objectives. "This study will examine the effects of increased water temperatures on the fecundity of females in one species of sea star, *Asterias indica*, as a model for understanding this issue in other tropical sea stars." The reader should finish reading the Introduction knowing what you plan to do and why it is relevant.

Literature review

Undertaking a thorough review of both seminal and current literature on your topic is critical and you will probably be asked to write up your review. (Proposals submitted for funding may or may not require a literature review section.) For the style of writing a comprehensive literature review, examine some of the journals that specialize in reviews (e.g., *Annual Review of Ecology, Evolution, and Systematics*) or some of the individual review papers that appear in some journals (e.g., *Biological Reviews*). Keep in mind the possibility that a well-written review on an interesting and important topic could become your first paper to submit to a journal.

Methods

This section is straightforward; but again, it forces you to think out the details of your project. The sequencing of methods is critical. They should parallel your objectives or goals. You may realize while you are writing Methods that your goals are unrealistic. Exactly what kind of equipment will I need? What will it cost? How am I going to select my study sites? How will I collect data on sea star egg production? Hmmm, I guess I need a boat ... that will mean boat safety training, and so on.

Analyses

As discussed above, with or without a formal research proposal, the experimental design and types of analytical or statistical analyses should be investigated prior to data collection. Statistical methods may change as the project develops, or as the data dictate, but it is important to have a plan so that you are confident your study design will provide you with the data you need for robust analyses.

Expected results or deliverables

It is good to think ahead to expected results and to put in writing what you think the contributions of your work may be. This exercise will allow you to evaluate whether your project has any practical applications. This section is often required by funding agencies, so it good experience to write it, even if it is not required by your university.

Timeline

Setting target dates for accomplishing key phases of your project will help you to evaluate whether your proposal is realistic. Your timing may be adjusted as things unfold, but thinking through all the elements of your project may save you from unexpected delays, or result in cutting back project goals.

Budget

You may not be involved with formal budget issues, especially if your work is funded by an external grant held by your advisor. It is still advisable to construct a budget that enumerates costs of materials, equipment, field and lab labor, transportation, and so on. This will be helpful to your advisor. Ultimately, budget constraints may dictate the scope of your project.

Literature cited

Compilation of literature cited for your proposal may be the foundation for future papers and a literature database.

Other elements

Your university may require you to include sections on how you deal with safety issues and the ethics of using animals or human subjects. You may want to include these even if they are not required because they deserve thoughtful consideration, if relevant.

Presenting your proposal

If preparing a proposal is required in your program, you will probably also have to give a proposal seminar that will be attended by faculty, students, and others who are interested in your topic. This is a great opportunity for you to get feedback, so be sure to leave plenty of time for questions and comments. It is common for the insights offered by the audience to improve, or even substantially change, your approach. Think of this as an "early warning system peer-review." Approaching your seminar experience with a positive attitude will make your presentation all the more engaging.

A proposal seminar is a good way to develop your mastery of giving oral presentations. Do a dry run with other students and your advisor and you will feel more comfortable giving your talk to a broader audience knowing it has passed muster. A proposal talk should follow the rules of talks prepared for professional meetings, as described in Chapter 7. For example, stick to the time limit, present clear visual aids organized in a logical fashion, and try to keep your eyes on the audience, not the computer or projection screen. If you know you will be nervous, have an image that gives a bullet-style outline of your talk, and by the time you get through that you should be more relaxed. Memorizing your opening sentence is also a good way to get started. Let the audience know why your project is important; how does it address some gap in our knowledge and contribute to conservation? The more detail you can give on proposed methods (including analyses and experimental design), the more valuable input you may receive, but time will constrain you from pursuing this in depth. It is perfectly acceptable at the end of your talk to say you are still not sure about the best way to do X or Z and to explicitly ask for suggestions.

The initial work you invest in designing your project will pay handsome dividends by boosting your confidence, increasing your efficiency, and improving the final product.

To summarize, the initial work you invest in designing your project will pay handsome dividends by boosting your confidence, increasing your efficiency, and improving the final product. It can take quite a long time to design a good project and many discussion sessions with your advisors, colleagues, and fellow students may be required. That said, most projects evolve over time and initial findings may well reset the kinds of issues that end up being the most interesting and valuable ones to address. This is normal.

EXECUTING A PROJECT

Now you have arrived at the foot of the mountain with a solid plan for your ascent. Probably the most exciting part of your journey lies just ahead. Enjoy this focused effort, because it may well be the only time in your career that you can tackle one issue in depth (in stark contrast to people with established careers, who typically have to juggle many projects simultaneously). This is also a golden opportunity to familiarize yourself with the process of managing a project. It is likely that someday, in whatever type of position you pursue, you will be required to procure funding and manage projects. Thus, familiarizing yourself with this foundation of the conservation world will be to your benefit in the long run.

Budget

Ask your advisor if you can see any proposals or budgets related to your project. You may not be allowed access to entire budgets because of sensitive issues such as salary levels, but it is reasonable to ask to understand parts that affect you directly. Be sure to find out when your funding expires and, if it expires before the expected life of the project, be clear on how the short-fall is to be accommodated.

If you are given some oversight of the budget, the next step is to create a spreadsheet that lists the funding categories (e.g., supplies, equipment, transportation, travel to conferences, and field or laboratory help) in which you record expenditures regularly. Keep copies of invoices because it is not uncommon for university accounting systems to make errors and you may be able to catch these.

Compare existing funding with your anticipated expenditures. Obviously, if funds fall short of what you will need, you will have to trim expenditures, reduce the scope of your project, or find more funding. There are diverse opportunities for securing additional funds to bolster your operational budget: small grants offered by professional organizations, foundations, government agencies, garden clubs, museums, and environmental groups as well as internal funds offered by the university. Funding opportunities can be discovered through simple Internet searches and scanning the websites of professional societies. Grants that you have been awarded, even for just a few hundred dollars, are a great addition to your résumé. In general, once you are successful in procuring funding, it becomes easier to get additional funds.

> *I had a particularly resourceful PhD student who raised over $20,000 during the 3 years of her field work by submitting 20 small research proposals (to the university, professional organizations, government agencies, and private foundations). She often recycled the same proposal to submit to funders so the process was not as time-consuming as it might appear. As a result, she was able to hire more help (which made her life more sane) and purchase better equipment. She also saved money through bartering – she worked in a national park and traded housing for her field workers for helping on another park project (a limited donation of field time in her area of expertise).*

Personnel: finding and supervising assistants

Finding assistants

Many projects would be enhanced by having more hands to make the work load lighter and to allow you to set more ambitious goals. If you have a budget that will allow you to hire assistants for a reasonable wage (most likely, but not exclusively, undergraduate students), then finding people should be relatively straightforward. If you have to seek volunteers, this will be a bigger challenge. In both cases you will want to be as selective as possible, hopefully as discriminating as you would be in choosing child care; your

If you are able to use assistants, be as selective as possible, hopefully as discriminating as you would be in choosing child care; your project is your baby.

thesis or project is your baby and the quality of your data will make or break your final product. Here are some tips for increasing your chances of finding competent assistants.

- **Advertise for help as soon as you can.** Don't wait until a few weeks before you need someone, because desperate employers do not make good choices. Create an attention-grabbing flier that highlights the most interesting and appealing aspects of your research. E-mail job announcements to all the undergraduates in your academic unit and any other units that may have interested students.

- **Take advantage of your contacts.** Ask faculty and graduate students for recommendations based on applicants' previous work or performance in classes, especially field or lab classes. Once you have applicants, check them out through this avenue as well. If you are a teaching assistant, keep your eyes open for promising student workers.

- **Seek students outside your university or discipline.** Students at other institutions, especially those without graduate programs, may welcome the opportunity to work on a conservation project. Send job announcements to relevant academic units at other institutions. If you can't find anyone with the specific skills you are looking for, consider seeking talented students from other disciplines. They learn quickly and may be anxious for a new experience; you might even convert a recruit to the conservation profession.

- **Consider volunteers.** If you do not have enough money to hire assistants, consider recruiting qualified volunteers. University students who are seeking their first conservation job may be willing to work for you in exchange for the experience, or for room and board if your work is in a remote location. If the work is straightforward, you may be able to recruit high school students or volunteers from environmental organizations. Some institutions provide matching funds for students who wish to work as research assistants. Volunteers can be very helpful, but they should still meet some baseline criteria. In other words, don't settle for an unqualified worker; they may cost you in the long run in terms of wasted time or compromised data quality.

Field ecology requires a lot of human power! I am a member of a large, interdisciplinary team that has actively integrated graduate, undergraduate, and high school students into our research program. Effectively integrating high school students into research programs can be challenging, but can be a positive and enriching experience for everyone involved. In our team, the graduate students have stated how mentoring high school students challenged them to become better teachers. Similarly, the undergraduate team members mentoring high school researchers have reflected how the experience has increased their feeling of responsibility within the team and challenged them to be better researchers. Furthermore, many of the students who have worked with us have indicated their summer research experience inspired them to pursue undergraduate studies in the natural sciences.

- **Conduct a personal interview.** Have a prepared list of key questions that you will ask all the applicants. You will want to ask them what types of experiences they are seeking in a job; about past employment (especially in related jobs); experience using any specialized equipment or techniques required by your project; experience entering data, and so on. During the interview, be clear about your expectations. For example, they may be required to work in unpleasant conditions (extreme temperatures, biting insects, physically intensive, etc.); work independently or as part of a team; and work until the tasks at hand are completed, rather than when the clock strikes 5 pm. Don't try to make the job sound easier than it is, especially considering the potential worker may not be used to your working conditions. In turn, provide time for them to ask you questions. Score them on a comparative rating sheet.

- **Test skills.** People may overstate their abilities to secure a position. For example, if you are looking for field help to interview loggers, conduct a mock interview to test their skills. Similarly, if your job requires physical fitness, don't simply ask if they are physically fit. Ask for examples of what they do to stay fit and for their experiences with other physically demanding jobs. Sometimes the opposite occurs and students downplay their talents. Of course, some people with no experience may be fast learners so you need to consider this possibility as well.

- **Check references.** Ask former employers and instructors about your candidates by phone or face to face. Do not be shy about directly asking what a person's shortcomings might be with respect to your needs.

- **Finally, trust your intuition.** If someone looks good on paper, yet you really feel uneasy about them after the interview, trust your feelings. Good chemistry is as important as good credentials.

Supervising

Be a good supervisor. It is incumbent upon you to make this a meaningful work experience for your assistants. It is often said that having help creates more work than it saves. There is some truth to this, but it is mostly in the start-up, training phase of the work and usually this investment pays off in the end. Communicate often and clearly about your expectations. If your assistants understand the project, they can troubleshoot and even give you advice on how things might be done more efficiently. Weekly project meetings allow people to give you feedback and provide you with an opportunity to keep the project on course. If you treat them as team members, not lackeys, you will create a mutually beneficial working environment.

Consider helping students to develop an independent project related to your work. Many students are looking for opportunities for an undergraduate honor's thesis or independent study as part of their undergraduate program.

Stress the importance of rigorous data collection. Think of conditions under which the worker is likely to be prone to sloppiness (gathering data in heavy rain or under attack from biting insects). On days such as this, reinforce this message and keep on reinforcing it. Explain why the data must be gathered rigorously. Regularly check the data and inquire about any irregularities. Share this experience with the workers.

Don't ask someone to do something you wouldn't do. Lead by example. Be reasonable in the hours you set; most people will accomplish the same amount of work in 8 hours as in 16 if you tell them that they can finish once a goal is reached. If you have a very long field day, arrange to give the student some compensation time on a less-demanding day.

Health, safety, and ethics

Before you begin your work, be certain that you have met all the special requirements of your institution. For example, if you work with vertebrate species or human subjects, you may need approval from an ethics committee. An increasing number of journals require proof of compliance before they will review a submitted article. You and your assistants may be required to have training in animal ethics, the ethics of work on human subjects, first aid, or the safe use of boats and motor vehicles. Check into this well before the field season as it can hold up your project if you are not in compliance. Keep a folder of your training and compliance documents and be sure your field assistants know what actions to take in an emergency. Who do they call first? What follow-up procedures are necessary?

> *I had a graduate student working on a controversial wetland mitigation project. He was interviewed by a reporter and photographs of him and his field assistants were published in the local paper along with their names. The following week I received a phone call from the head of our Institutional Animal Care and Use Committee. She had seen the newspaper and did not recognize one of the assistant's names as having received training. It turned out the assistant was a volunteer for the day and was restricted to recording data, not handling frogs, and so was not in violation. It pays to be in compliance!*

Ethical considerations are not restricted to meeting university-imposed standards. Whether you are studying plants, animals, or human subjects, you should be sensitive to cultural issues and the concerns of local people, especially if you are working in an area where you are an outsider. While conducting field research, be sensitive to cultural issues and the concerns of local people. For example, on both public and private property, it is common courtesy to secure permissions and make your research goals clear to property owners and managers. If possible, at some level, engage local students or stakeholders in your research. If your research is in another

While conducting field research, be sensitive to cultural issues and the concerns of local people.

country, you may in fact need special permissions and to be affiliated with a host institution. In short, be respectful and in the long run your project and all involved will benefit. We will return to this issue in Chapter 10, Making a difference.

Equipment

As a project manager, you should have a formalized system for maintaining your equipment. It will save you time, money, and stress if you take the time to do this systematically. Your assistants should be familiar with the system as well. This will highlight to them that it is important to care about the condition of your project tools. Keep an inventory sheet that notes the specifics of the equipment, its maintenance status (e.g., for a pH meter, when was the probe last replaced?), and its location. Have a separate sign-out sheet for equipment that is used in various places by different people. Keep a folder with user manuals in one location. This accounting system will also make it easier for your advisor to locate and use the equipment in a subsequent project.

Managing your database

The importance of managing a database is so obvious that it should not require emphasis, but our experience has convinced us not to take anything for granted. There are countless examples of students who have lost hand-written field data sheets that were not copied, electronic data files that were not backed-up, or have not entered the first year's data until the third year and have forgotten the coding system used in that first year, rendering the data meaningless. Some tips for guarding your data are given here.

- **Take clear and thorough notes on data-collection methods and field observations.** You can always archive excess notes but you cannot easily recall data you haven't collected.
- **Develop data-collection sheets in phases.** You may anticipate work progressing one way and when you test the data sheet, you may find you need to change it.

- **Find a spreadsheet program that best accommodates your data needs.**
  Look for one that will allow you to interface easily with statistical programs.

- **Research and select a data recording method in advance.** Techniques will
  vary from hand-held computers to field hand-written data sheets.

Graduate students commonly report that they are most comfortable with
a combination of data archiving techniques. "I prefer field data to be hand
written, but photographed (with a digital camera) or scanned in at the
end of the day as well as photocopied (so at the end of the day I have
3 copies). I really liked having data entered into a spreadsheet on 'the
cloud' as soon as possible. I used Google Drive which let multiple people
enter data into the same spreadsheet simultaneously and constantly was
'backed-up.' As soon as a field tech wasn't working for me anymore, I
disabled their ability to edit and view the data on the cloud."

- **Establish a protocol for keeping track of data collected by field assistants.** Check it regularly.

- **Back-up, back-up, back-up.** Do this often. Do not keep backed-up files in
  the same location as the originals. Spread the wealth.

A master's student in New Zealand scrupulously maintained back-up copies of all data files, analyses, and drafts of thesis chapters and always stored these in at least two different locations. However, just before printing the final version of her thesis, she moved away from her university, placing her computer and all back-up files in her car for the short move. The car was left unattended for less than 30 minutes and the computer and all back-up files were stolen. Two years of hard work had apparently vanished. Happily, the thief threw the back-up disks in a river and the files from one waterlogged disk were recoverable. Moral: the potential for disaster is always there and there is no such thing as too much paranoia when it comes to maintaining back-ups.

The initial period

Many students find the first days or weeks of a project the most stressful period during their research. You tend to believe that if all does not go well from the beginning, your project will fail. Your research experience will be much more enjoyable if you realize that the initial phases of your research, no matter how well planned they appear on paper, often have a pilot phase during which you will need to modify your methods or adjust to unexpected bumps in the road. Laboratory equipment may malfunction. Weather, plants, animals, and human subjects do not always behave according to plan. Study sites may be bulldozed. Access to archival information may be blocked. If you assume every data point will make or break your project, you will generate lots of stress and you may be blind to assessing things that should be abandoned and new opportunities for exploration.

A research design was completed and approved for studying conflict resolution between zapovedniks (Russian nature reserves) and economic development agencies. The design was based on significant preliminary work, including many interviews with political officials and staff from government agencies and environmental groups. Several months into the

study, two things happened. First, zapovednik management was moved to the agency responsible for forest management (exploitation), undermining the structure of the study, which was based upon existing agency relationships. Second, a significant volume of new documents, previously housed in inaccessible archives, were found. These documents involved agencies that were not included in the original research design. Even with good planning events can occur that undermine a project.

Figure out the most efficient way to collect the data that is key to researching your question and do not waste time and energy on intensive data collection that does not pay off. Whenever you change methods, you have to think carefully about the comparability of data gathered before or after the change. In some cases you may need to discard the earlier data, but this is probably preferable to going forward with a flawed approach.

For the first field season of my dissertation, I planned a study on egg mass variation within two genetically different types of female salamander. I planned a breeding experiment in which I would put pairs of salamanders in breeding chambers and measure the resulting eggs. Then nature threw me a curve ball. I captured hundreds of females but

only two males. Without being able to pair the salamanders the breeding experiment was a flop. The experience, however, revealed a novel and much more interesting question: how were the salamanders reproducing with so few males? I am now returning with new interests and new methods to address this enigma.

Advisor participation

Advisor involvement in your project may range from benign neglect to micromanagement. Hopefully, you are able to work out a relationship somewhere between the endpoints of this continuum. If your advisor errs toward neglect (or perhaps supreme confidence in you), we recommend that you spend some significant time in the field or lab with your advisor early in the project. This has three major benefits.

1. It gives your advisor an appreciation for the conditions you are working under and possible challenges that you may face.
2. It allows the advisor to see whether what seemed clear on paper is actually what is being done in practice. Having an experienced person observing your operation may help identify more efficient procedures or eliminate data collection that is unnecessary.
3. Being on the scene is likely to stimulate your advisor to think of new opportunities to improve your project. For example, perhaps adding one easily measured parameter to your data collection will allow you to address a new and important question.

> *I extended my graduate research by failing to ask my advisor to visit me in the field until the end of the first field season. Standing in a cattail marsh, up to our armpits in water, I was proudly demonstrating experiments on red-winged blackbird nests that I had thought of since my last committee meeting back on campus. The treatment results were amazing and I was waiting to hear praise from him when, from across the marsh, came a simple question, "Where are your controls?" All the data I had collected had come from treatments. The hours spent in those marshes during my first field season may not have produced publishable data, but the lesson I learned that day was worth far more than a publication.*
>
> *My advisor taught me another lesson that day, as important as remembering to balance treatments with controls. Graduate advisors serve as mentors, and when a student makes a mistake, that is not the time to chastise them. Years later, now a professor, I was visiting one of my students who was conducting field experiments on nesting birds in North Dakota grasslands. After visiting scores of treatment nests I asked to see some of the control nests. She replied that since she hadn't found any differences between early treatments and controls, and because controls were taking a great deal of her time, she had discontinued visiting the controls. Remembering the grace with which my advisor had treated me years ago, I suggested we go have a cup of tea and revisit her study design.*

After the first round of data gathering

You may be inclined to say, "Phew, now I can rest," but after a short period of recuperation you need to be hard at work again. Although you may have analyzed dummy data sets and completed pilot studies, there are some key steps to take after your first data-gathering session to ensure project success. In addition to constructing your database and carefully checking data accuracy, it is extremely important to run a preliminary analysis of your data. Be sure to start with your planned analyses and don't make a so-called Type 3 error – that is, only checking data when the results seem counterintuitive.

A preliminary data analysis can lead to some important outcomes.

- Some of your data sets may be adequate to answer a subset of your research questions. If this is the case, you can begin the process of writing early, possibly even generating a publishable article (see Chapter 8, Writing papers).

- Alternatively, preliminary data analyses may suggest that more data or more replication within your existing survey or experimental design are required to test the questions you have posed. If you have a field-based project, you may be able to assess how many more field seasons you will need to collect enough data to answer your research question or complete your project.

- Preliminary analyses may suggest that you abandon a particular line of study altogether. It is far better that you find this out early in your program rather than attempt to salvage a train wreck at the end.

- Your data analyses may reveal some promising but unanticipated findings, suggesting a new line of study that builds on the first data set.

It is natural for the trajectory of a project to evolve and change over time. In fact, it would be rare and very surprising if it did not. The key point is that it is vitally important to assess progress early to minimize the risk that the entire project is an unmitigated disaster. Your program will be dynamic and may lead to exciting routes that you did not anticipate at the outset; you will not know this, however, if you do not track your progress regularly. Continuously discussing your work with your advisors, other students, and various conservation professionals will also lead to improvements. Similarly, keeping abreast of the relevant literature will provide a font of new ideas.

In addition to analyzing data and discussing your work, the intervals between periods of data gathering are often the best times to deal with administrative tasks such as submitting progress reports to funding agencies; filing reports on safety, health, ethics, and personnel; refurbishing equipment; and so on. Keeping up with these commitments is not particularly rewarding, but it is useful experience for future employment that will almost certainly include these kinds of responsibilities.

It is natural for the trajectory of a project to evolve and change over time.

NON-COMPLETION

Students withdraw from their degree programs for a broad array of reasons. Some discover that research or executing a major project is just not for them. Still others decide that family pressures or financial constraints make continued education inappropriate for them. There is nothing shameful about withdrawing from a program; many people fail to complete the course they started. If you do decide to withdraw, be sure to discuss your decision with your advisor long before you depart, preferably weeks before. It is your responsibility to leave the project as intact as possible. This will facilitate finding a successor, wrapping up the project, or whatever is deemed best. Even if one of the reasons you are leaving is that you have a poor relationship with your advisor, there is no need to make a bad situation worse. Think about ways to exit with your head held high and explain your decision to the chair of your academic unit.

One of the toughest meetings I recall as an advisor was with a student who broke down in my office after deciding to withdraw from his doctoral program. He was devastated that people would think he was a failure and had no ability to see a task through to completion. It was hard to know what to say that was not trite or condescending, but I think that he really just needed to be reassured that many people before him had decided to follow paths other than a graduate program and that many of them had made important contributions to conservation in other ways.

WRITING A THESIS OR FINAL REPORT

Writing is both an acquired skill and an innate talent. If you plan to make your living publishing novels or books, it certainly helps to have talent. However, writing is also a skill, and with practice and patience most people can become competent writers. Hopefully, you have had some practice with this if you wrote a proposal or had a technical-writing class. There are many user-friendly books available on writing scientific papers that are worth perusing.[6.4] We also suggest that you read a variety of journal papers in your field with an eye toward the written presentation. Just as language skills (grammar, syntax) are learned by children whom their elders read aloud to, the anatomy of a professional paper will seep into your psyche if you actually study published papers. In other words, language is catching – you just need extended exposure and practice. Here we present a few tips on the special process of thesis writing; we have devoted an entire chapter to writing journal papers and the overall process of writing is covered in more depth there (see Chapter 8). There are many similarities between writing a thesis and a final project report that will make this section relevant to students in non-thesis programs and other conservation professionals.

Thesis or final report preparation is often the most intense and stressful phase of a degree program. There is usually an impending deadline – such as the time when funding runs out or when a candidate has accepted a job and is desperately trying to finish before starting a new work commitment. Do not underestimate the time needed for this final bout of writing; it can take many months, even if you prepare chapters as concise papers for scientific journals. If you build writing time into your project from the beginning, you may avoid the stress of funding drying up before you have completed your degree. Here are some constructive steps to minimize your stress and maximize your enjoyment of reporting your work.

Do not underestimate the time needed to write your thesis or final report.

Write about your work in phases. You can write the description of study sites, methods, and analyses for each phase of your project as it is completed. In fact, it is easier to recall your methods while the work is being done than to try to reconstruct them from notes.

Get early feedback on your work from your advisor and other students so they can spot major problems in time to rectify them. This is essential as the quality of writing improves over time. To mix metaphors, each draft you write is like peeling away the layers of an onion, eventually revealing the pearl hidden inside. It is particularly

Get early feedback on your writing so that major problems can be rectified.

important to submit chapters to your advisor as they are completed. There is nothing worse for an advisor than to have a 300-page document dumped on their desk, having seen none of the written material previously. (That said, some advisors may want a completed product so it is always best to ask ahead of time.)

If you have a committee, send them chapters to review only after they have been reviewed and approved for submission by your advisor. They may prefer to see one chapter at a time or the entire document. In either case, give them an estimated date of arrival in advance. If committee members receive a draft that is in good shape, they will have more time and energy to provide input on the scholarship, rather than spending their effort on your writing and organization.

Producing a final document can be challenging and it is essential to stay as focused on the task as possible and to limit the number of distractions. Many students see other (often peripheral) angles to their work and think about additional chapters or papers they could write. Whereas this is potentially interesting, it can divert you from the primary task. Other papers that are not essential to the core themes of your project should be put on a back burner until you have submitted. You can tackle them at a later date, or indeed the topic may be so interesting that you can pursue it in more depth in the context of your next position.

While it is necessary to stay focused on the major writing tasks at the end of your program, some distractions are important to remain sane and fresh. Most conservation professionals work in this discipline because they love the environment. Take some time to enjoy the outdoors; it can help remind you of your ultimate goals.

Outlets for your work will vary with your target audience and are ultimately most effective if pitched to broad audiences through many outlets. Writing your chapters as scientific papers for journals is a critical step in communicating your work to the academic community. However, you may wish to reach other audiences through popular literature, web-based communications, and oral presentations. Ways to do this are discussed in more detail in Chapter 8. For example, you should also plan to give seminars on key findings from your work in your department, at conferences, or preferably both; Chapter 7 gives some guidance on this.

FINAL DEFENSE

The nature of a final defense varies with programs. Some require a public seminar followed by an oral examination by your advisory committee. Other defenses may involve a panel of outside reviewers who examine the final document and an oral examination may or may not be a part of this process. This is an exciting and intimidating time for most students. You have to realize that you know more about your body of work than anyone else and you should enjoy sharing it. On the other hand, in the grand scheme of things, you still have a lot to learn and you may receive an ample dose of humility from your examiners. Oral examinations are often used to test the depth and breadth of your knowledge. Examiners will plumb the limits of your knowledge and may well exceed it, so don't be afraid to say, "I don't know." Keep your sense of humor and remember that examiners may use this time to fluff-up their feathers as well. It is not uncommon for an examination to evolve into a debate among your examiners in which you become a bystander for a short period.

What you learn from your defense may expand your horizons and help you to keep your research in perspective. Feedback from this final defense may lead to improvements in publications from your work and it may be the last time in your career you have the undivided attention of a panel of professionals.

Further readings and notes

6.1 Peters, R.H. 1991. *A Critique for Ecology.* Cambridge University Press, Cambridge.

6.2 Hunter, Jr, M.L. 1989. Aardvarks and Arcadia: two principles of wildlife research. *Wildlife Society Bulletin* **17**, 350–1.

6.3 Driscoll, D. and D.B. Lindenmayer. 2012. Framework to improve the application of theory in ecology and conservation. *Ecological Monographs* **82**, 129–47.

6.4 See note 8.1.

7 Attending conferences and making presentations

Are you a party animal? Someone who is always open to a few days of hedonistic pleasures? Well, the bad news is that you may find your typical conservation conference a fairly dull affair, at least compared to the "anything-goes" business conventions depicted in popular culture. For example, you won't find many open bars sponsored by corporations trying to court the goodwill of the conservation community with free alcohol. The good news is that conferences are a very valuable experience, rich with opportunities to learn about your profession and to make connections with far-flung colleagues and prospective employers, while having fun too. For many people they are the highlight of their professional year. In an era of increased video-conferencing capabilities, conferences provide an uncommon opportunity for people with shared interests to gather in the same place over an extended period and really connect with each other. That said, attending conferences is often quite expensive, in terms of both time and money, and they often come with the responsibility of developing a talk or poster, so before turning to the mechanics of conference attendance and presentations we need to consider how to select a conference.

WHICH TO ATTEND?

Across the breadth of conservation there are certainly hundreds, perhaps thousands, of professional gatherings each year, ranging from huge meetings that attract thousands of people from around the globe to narrow symposia and workshops attended by a few dozen specialists. Some of these meetings are

Saving the Earth as a Career: Advice on Becoming a Conservation Professional, Second Edition.
Malcolm L. Hunter, Jr., David B. Lindenmayer, and Aram J. K. Calhoun.
© 2016 John Wiley & Sons, Ltd. Published 2016 by John Wiley & Sons, Ltd.

one-off events; some are annual gatherings of professional societies; a few, like the congresses of the World Conservation Union, are held every 3 or 4 years. How to choose among so many alternatives?

Topical relevance

Naturally, the first issue to address is relevance: if your focus is on carbon sequestration legislation then a wonderful conference on seabird conservation will not be a good investment of limited resources. However, sometimes relevance is harder to sort out. How do you choose between a broad-based gathering like the biennial congresses of the Society for Conservation Biology, where there might be several sessions on environmental legislation spread over 4 days, versus a small, 2-day meeting that attracts only policy specialists and environmental lawyers? The former will provide a more diverse experience that will enhance your development as a conservation professional writ large; the latter will be more directly applicable to your work. There is no simple answer and probably other issues will need to be considered too, as we will see below.

Limited resources

Financial limitations and the press of time keep most conservation professionals from attending as many conferences as they would like to in a world without constraints. Of course, these constraints will loom large for students too and you are likely to be limited to attending one conference per year, and

Limited time and finances make selecting conferences to attend an important decision.

perhaps just one during your degree program. The cost of attending a major conference will probably be dominated by an air fare, especially because professional societies usually encourage student attendance at their conferences with subsidized registration costs and cheap options for accommodations like dormitory rooms. To reduce transportation costs, you may be able to find a regional conference that is occurring within a day's drive of your university and share the costs with a whole group of students. Many universities will happily cover the cost of a vehicle if several people are the beneficiaries. Regional conferences are also likely to be cheaper than an international conference because they tend to be shorter and are more likely to be on a university campus than in a big city conference center.

Many conferences are scheduled during the summer to avoid conflicts with the academic calendar, but unfortunately this is also a prime season for gathering research data. Slipping away for a day or two may be one thing, but a major conference can eat up a week, not including the days, perhaps weeks, it may take you to prepare a presentation.

If time and money constrain you to attending only one major conference, it is probably wise to wait until you near the end of your program. At that time you will be able to give a complete account of your work and use that as a springboard for finding your next opportunity. You are also much more likely to find financial support for attending a conference if you are ready to give a presentation on your research. Hopefully you can also attend some conferences earlier in your program, perhaps presenting a poster on your research design, and thus solicit early feedback and network with people doing analogous work. The bottom line is that limited time and finances make selecting conferences to attend an important decision.

Conference type

Conferences come in many flavors that may be more or less appealing. Many people like to attend the annual conferences sponsored by a particular professional society because going to the same conference year after year is a great way to develop a network of friends who are following the same pattern. Alternatively, by going to a Society for Conservation Biology meeting this year, one held by the International Society for Ecological Economics next year, and then a conference focused on a particular type of ecosystem, you will diversify your experiences. Others prefer to seek intimate meetings, such as those that are confined to a specialized subject or those that cover a wide topical range but are attended only by people from a small area. It is certainly easier to have repeated, meaningful conversations at a meeting where you are seeing the same people at every session and meal.

Some students prefer to attend student-only conferences (e.g., the Student Conference on Conservation Science held each March at the University of Cambridge, UK, and at other locations)[7.1] because they feel more at ease in an all-student gathering. Locally organized student conferences are a great way to minimize cost issues and to give students practice at presenting without the pressure of a large assembly. Commonly, all the conservation students at universities within a day's drive are invited to assemble at one university on Friday evening, have sessions on Saturday and Sunday morning, and return home on Sunday afternoon. In some larger programs, student conferences are held entirely in-house and offer a great way to keep abreast of one another's work and practice giving talks to an audience of friends.

I started out going to a different conference every year, and while it was interesting to see the differences I sometimes found it a bit lonely, especially at the large conferences, because I'm not very good at initiating conversations with people I don't know. After a few years I decided to focus on attending Society for Conservation Biology conferences and soon I had a group of friends that I was seeing year after year.

My first conference presentation was at our state's Association of Wetland Scientists, a small group that was really supportive. They were a perfect audience for getting my feet wet.

So how do you systematically sort out all the issues and options? Again, there are no easy answers, but you can start with a conversation with your advisor.

Pat: "Could I talk to you soon about going to a conference?"

Terry: "I've got 10 minutes right now."

Pat: "Okay, I was wondering if there is enough money for an air fare to attend the SCB conference next June. I can get my registration cost covered by the university, and my cousin lives there so my housing and some meals would be covered."

Terry: "I'm glad you've given some thought to logistics but the air fare will probably be almost $1,500 and taking almost a week off during your field season would be difficult. Next year the SCB conference will be much closer and cheaper to attend. More importantly, you will be in a position to really tell your whole story by then."

Pat: "But Jamie is going to be my field assistant again this summer so we could keep the surveys going."

Terry: "Let me check the budget but I doubt it will work, especially since you plan to hire two assistants for next summer. Actually, a better idea might be to present your work at a workshop on population monitoring that the Department of Conservation is organizing for next October."

Pat: "Okay, that's certainly better than nothing."

Terry: "It's a lot better than nothing; these workshops are often a better place to meet people, especially if you're still thinking you'd like to work for the DOC eventually. And don't forget that grad student symposium at Alpha University next month. That's really the best place for presenting a progress report on your work."

CONFERENCE INFORMATION

The best place to learn about upcoming conferences is through professional societies.[7.2] They sponsor many conferences themselves and their websites and social media hubs often list a wide variety of conferences organized by other groups such as government agencies and environmental NGOs. Start your search early and pay particular attention to timing issues. Deadlines for submitting papers are usually many months before the conference. For some conferences you can register at the door, but others require preregistration and may reach a capacity limit long before the conference. It is also common to encourage preregistration with a lower conference fee. If you want to present a paper, check out the opportunities carefully: some conferences are wide open; some only allow invited presentations; some have a limited number of openings for which the competition may be very easy or very difficult.

ATTENDING TALKS AND OTHER SESSIONS

If you are lucky, your presentation will happen early in the conference and you can relax in the afterglow of your moment in the sun, turning your full attention to other presentations rather than worrying about your own. Many conferences have from two to ten or more simultaneous sessions, so the major issue is deciding which talks to attend. Newcomers to conferences often spend considerable time poring over the program to develop an elaborate game plan for skipping from session to session to hear all the talks that are closest to their interests. In contrast, experienced conferees often take a more relaxed approach, choosing a single session and sticking with it until a refreshment break. One problem with the skipping-around approach is that you are likely to arrive after a talk has begun and find yourself squeezed into a crowded corner by the door where you cannot see the visual aids, and you will miss most of the question-and-answer sessions that follow talks.

> *Over the years some of the best talks I have attended are ones that I never would have picked out of the program, but I heard them because I could not be bothered to go racing from room to room. I recall a fascinating talk on conservation of beach invertebrates in Poland that I never would have encountered if I focused on talks about my specialty.*

If you have a question for a speaker, you may have a chance to ask it at the end of the paper, but time is often limited and you may need to wait until later. You can probably have a more meaningful interaction later anyway compared to a question shouted across a room. If you familiarize yourself with the speaker's appearance and location, you can approach them at the end of the session and talk to them then or make plans to meet later. Asking good questions, whether in the midst of a session, or later while talking to a person one on one, is a skill to develop. Pay attention to the questions asked by other audience members and you will soon learn how to ask a question that is concise, direct, clear, and respectful.

Many conferences have sessions that are analogous to paper sessions – lots of people talking about conservation in a structured format – but rather different in terms of their goals and organization. These go by various names – workshops, discussion groups, short courses, and others – that may or may not accurately reflect their structure. For example, a discussion group might consist largely of listening to an appointed panel with little room for audience input, whereas a workshop may be an exercise that entirely revolves around group participation. Some of these sessions are loose analogues of speed-dating, in which a speaker has just a few minutes to talk, or attendees move around a room interacting with each other in very short periods. Our basic advice is that if the topic holds any interest for you at all, check out the alternative formats. These are often more intimate sessions where you are likely to find people with similar interests, and if you are unsure about how the format works, ask a veteran.

NETWORKING

One of the most important functions of conferences is almost purely social: developing relationships with new colleagues and reaffirming old relationships.

One of the most important functions of conferences is developing relationships with new colleagues and reaffirming old relationships.

There are many venues for this, ranging from the simple coffee and lunch breaks that occur at all conferences to the elaborate receptions and banquets that usually begin and end major conferences. As three people whose natural inclination is to hide in the corner at large social gatherings, we are no experts at how to thrive in these settings. However, we have a few suggestions and observations that go beyond the sort of basics your parents taught you, such as don't curse and don't talk with your mouth full of food. Here they are, in no particular order.

- The larger the gathering, the harder it is to have a meaningful conversation because of noise and interruptions. If you really want to talk to someone try to find a more relaxed setting, such as sharing a meal where people are confined to a table for an extended period. (You may recall from the Preface that this book was conceived while we sat outdoors at a conference.) Field trips are also an excellent opportunity for socializing and some conferences have events designed to foster student interactions.

- Location matters. At most conference meals it is hard to talk to more than the persons on your left and right and perhaps across the table, so choose your seat carefully.

- Avoid spending too much time with people you already know very well, especially people from your university that you see almost every day during the rest of the year.

- Missing sessions is not analogous to missing classes. If you initiate a useful conversation during the coffee break (e.g., discussing a collaboration), it is almost certainly more important than running off to hear another paper.

- Although it may be more fun to chat with other students, don't be reticent about interacting with established professionals. They are much more likely to be useful contacts for advancing your career. Even the old luminaries of conservation are more approachable than you probably realize, because anyone who cares about conservation cares about nurturing the next generation of conservation professionals. If you met someone at a previous conference, introduce yourself again. They may not remember you (if they are luminaries they are probably old enough to suffer from neuron rust) but they are much more likely to recall you after a second chat.

- If you are reluctant to introduce yourself to one of the leaders in your area of focus, ask your advisor to introduce you. This can even be done *in absentia*: "Hi, my name is Pat Lake, and my advisor, Terry Black, asked me to say hello. I've always admired your work and wanted to meet you." You would be very surprised to know how many established professionals are a bit shy themselves, and will genuinely welcome a conversation with you.

- At most conferences, a sizable portion of the people are lonely but are too shy to do much about it. Take your courage in hand and introduce yourself to strangers. Invite yourself to tag along with a group going to dinner. Track down a speaker you enjoyed at the next refreshment break. You know you share a common interest in conservation so it's not like approaching strangers in a bar.

- Consider having a business card, even if it is one you print yourself, to give to people. It's much more convenient than scribbling your email address on a scrap of paper that is likely to be lost.

I was standing with one of the founding fathers of conservation biology when he was approached by a student who wanted to shake his hand. He responded by shaking the young person's hand, smiling broadly, and saying, "It's a privilege to meet you."

PROFESSIONAL-SOCIETY ACTIVITIES

Many conferences represent a golden opportunity to focus on a small but significant part of being a conservation professional: your membership in a professional society. There are hundreds of professional societies that attract conservation professionals to their ranks and most of the larger ones sponsor at least one conference per year.[7.2] Professional societies are best known for publishing journals and sponsoring conferences, but they

Professional societies will be a critical vehicle for professional development throughout your career, and the sooner you get on board the better.

undertake a diverse array of activities and at a conference much of that action can be experienced directly. For example, sometimes there are debates on how the society should become involved in various conservation issues. Here you can witness the perennial tension between members who want to use the credibility of the society to weigh in on a conservation topic and members who believe that a professional society should only be a source of objective information, lest it be mistaken for an environmental advocacy group.

Although involvement in policy issues is always interesting, this is actually a small portion of what professional societies do. At their foundation these societies work to further conservation by providing mechanisms for exchanging the latest knowledge and perspectives, thereby continuously improving the capabilities of conservation professionals. In other words, they are likely to represent your chief process for professional development throughout your career, and the sooner you get on board the better. Indeed, given the strongly subsidized student memberships that most societies offer, you should consider finding room in your budget to join a number of professional societies – and remember to add these to your résumé.

> *I started with the Ecological Society of America and The Wildlife Society while I was an undergraduate and added a couple others during graduate school. When the Society for Conservation Biology came along I jumped on that bandwagon and that's where most of my attention lies now, but I still belong to four other societies.*

Conferences are a stew pot for a society's governance activities, typically with a general membership meeting plus meetings of various boards and committees, most of which are open to observers. If you find that the mission of a particular society really resonates with you it is never too soon to volunteer to join a committee.

> *One of my PhD students was elected President of our state chapter of The Wildlife Society. I was worried about how this might conflict with time allocated to his research work, but it was a unique opportunity for him to develop leadership skills working with a large group of established professionals, so I gave it my blessing.*

This may be an unusual case, but the general point is that you can become involved in professional society activities now and they represent a chance to hone some skills and develop some connections that will serve you well in the future. And your first chance to do this is likely to come while attending a conference.

PRESENTATIONS

Attending a conference often comes with a major responsibility (and source of excitement or anxiety): presenting your research work in either poster or oral format. In some smaller conferences you can simply volunteer to make a presentation and you will be assigned a time and location. More commonly there is a formal selection process that starts many months before the conference with the submission of an abstract.

Submitting an abstract

No doubt you have read scores of abstracts for journal papers, but this might be your first experience at writing one. They are not easy. Distilling the essence of your work into a single paragraph that tells a comprehensive and cogent story is a challenge. As the French mathematician Pascal once wrote: "I have made this letter longer than usual, only because I have not had time to make it shorter." Here are the essential elements of writing a good conservation science abstract, presented, in the spirit of conciseness, in four sentences. Develop a title that is informative and interesting to help attract an audience. Follow the standard format of a scientific paper with Introduction, Methods, Results, and Discussion each condensed to one to three sentences (see Chapter 8). Tell a coherent story that culminates in your main conclusion, hopefully one with direct relevance to conservation, not just a call for further research. Provide critical evidence to support your conclusion, preferably with specific, quantitative results.

Distilling your work into an abstract that tells a comprehensive and cogent story is a challenge.

With these guidelines in mind, read some abstracts from a major journal or from a conference (many professional societies archive these on their websites).

You will find many abstracts that do not follow these guidelines very well. Some of this occurs because certain topics do not fit well within these guidelines (e.g., a discussion of how wetland regulation policies are evolving will probably not have Methods and Results). Some of this occurs because it is hard to write good abstracts and many people do not take enough time to do it well. When writing abstracts for conferences there is often a gap of several months between writing the abstract and making the presentation, and this can lead to abstracts that conclude with open-ended promises such as "Results will be discussed in the context of reforming endangered species legislation." It is better to tell the story as you know it now, and if it evolves during the interval between abstract submission and the conference, which is likely, you can still tell the new version.

Why bother spending a couple hours writing a paragraph? First, many conferences have a finite capacity for presentations and thus a poor abstract may lead to outright rejection. For example, less than 20% of the abstracts submitted for the Student Conference on Conservation Science at the University of Cambridge are accepted. Similarly, the capacity for oral presentations is often especially tight, so an abstract submitted for an oral presentation slot may be redirected to a poster session. Second, many conferees read abstracts to decide which presentations to attend. Third, professional societies often have student presentation competitions and abstracts are the first filter in this process. There is more on this below.

The format for submitting an abstract is usually very rigid, with a maximum length, a particular style for reporting your affiliation, and so on. Follow the instructions exactly. Improperly formatted abstracts are a huge headache for conference organizers and submissions are sometimes rejected on this basis alone. You may be asked to designate a topical area for your submission. Examining archived conference proceedings may give you some idea of the types of presentation that are likely to precede and follow yours if you choose "stream ecology" or "fish conservation" or "water policy."

When you submit your abstract you will probably have to indicate whether you would prefer to give an oral presentation or a poster. Most established professionals prefer to give oral presentations because: (1) they are somewhat more prestigious (due to the limited capacity for them at most conferences); (2) they often attract more attention (as measured by person-hours of viewers); and (3) they are a bit less time-consuming to prepare, at least for seasoned presenters. On the other hand, many people, especially newcomers, prefer poster presentations because it is easier to have a conversation with a few people than to talk to a large audience. It may also be more productive in terms of having real dialogue with someone who is keenly interested in your work and can offer personalized feedback. A poster presentation is probably a better choice if this is your first conference or if you are reporting on a research project that is only partly completed.

Poster presentations

If you have never been to a poster session, imagine a marketplace with dozens of merchants trying to hawk their wares to hundreds of potential customers in a single crowded room. In this case, the merchants are very demure; they would not dream of loudly exhorting the crowd to gather at their poster. They just stand beside their poster as people wander around, hoping that their poster will catch the eye of an interested person. When people stop, a conversation will probably ensue, perhaps for only a minute or two, perhaps much longer.

In this setting, the key to brisk sales is an eye-catching poster and this probably means both an intriguing title and one or two large, attractive images ... literally attractive.[7.3] It is possible to go overboard with a cute title, but certainly almost anything would be better than: "Life history of the alpine lily, an endangered species" or

Posters need to attract your audience and convey a few key ideas.

"Visitor attitudes toward conservation at Alpine National Park." The first requirement for an attractive image is that it be large enough to be clearly visible from 2 or 3 meters away. Secondly, it should attempt to engage the viewer and make them want to find out more: What are those people doing? What kind of creature is that?

Once you have a person's attention you can initiate a conversation with a question such as, "Would you like to hear about my work?" and if the answer is "yes" then you have a few minutes to deliver your message, what communication experts often call an "elevator speech." You should probably aim for about 3–5 minutes, but it may easily extend longer if you are asked questions, a good sign that the person is genuinely interested and not just being polite. During your mini-presentation the main function of the poster will be to provide some key figures or tables you need to display your results and perhaps a map to show where you work. Figures are usually preferable to tables, especially because they are easier to interpret from a modest distance, but sometimes a table will work better.

If the key elements of a poster are a title and image to attract a listener and then some figures to use while talking about your results, then why is the typical poster cluttered with lots of text material? There are two answers: first, some posters are designed to be read while the presenter is absent; second, most posters try to present far more material than can be realistically absorbed by the average reader in the time they will spend standing before a poster. The first issue is easily resolved by determining the structure of the poster session; if they are on display for an extended period then that calls for a different design than if they will be visible for only an hour or two when you are present. If the poster needs to stand alone, you will need some bare-bones text to lead readers through the main parts of your paper: an Introduction to provide some context, Methods to describe what you did, and so on, perhaps 200–500 words at most. The mistake people make most often is overwhelming their reader. Surrounded by scores of other posters that are competing for attention it will be unusual for anyone to spend more than 5 minutes reading your poster, unless the topic particularly interests them. First, you need to ask yourself, how much material can I actually convey to a reader in 5 minutes? How can I strip my story down to the essential elements to be convincing and memorable? Then you need to organize that material in a visually accessible design. This means large, simple fonts, an outline format based on key phrases, a visual design that facilitates flow through the sections of your poster, and, most important of all, avoiding excessive text.

Imagine standing in front of a poster reading the preceding paragraph compared with seeing its essence distilled like this:

Why are posters cluttered?

- Designed for reading, not presenting

- Provide too much material

Solutions

- Limit material to 5 minutes' worth

- Use an outline format

Sure, some nuance is lost in the latter version, but which one will you actually take the time to read in the midst of a poster session and which one is more likely to stick with you?

Aim to have a first draft of your poster done a couple weeks before the conference so that your advisor and others can review it for you. Some fresh sets of eyes will almost certainly identify opportunities for major improvements.

For people who are really interested in your work, it can be useful to have a short handout, say both sides of a sheet of paper, that contains your abstract and key figures. These can be left in a large envelope hanging from the poster if it is unattended at times. Of course, this handout should contain your contact information so that the interaction can continue by phone or e-mail after the conference. Conversely, you should be prepared to collect contact information for people that you want to talk to later. Obviously face-to-face interactions are best, and you might want to suggest meeting for a meal later in the conference if you think this would be fruitful.

Many first-time poster presenters are disappointed with the volume of inter- actions; you are likely to spend more time standing alone than engaged in animated conversation. This is not a reflection of the quality of your work. The reality is that poster sessions are social events and many participants will be more focused on chatting with colleagues than viewing posters. One way to avoid standing alone is to interact with other nearby poster presenters. Ask them to describe their posters and they will almost certainly reciprocate.

Oral presentations

Some people love being in the limelight of public speaking; some people con- sider it to be a particularly excruciating form of torture. Wherever you fall on this continuum, it is an important skill to hone because some public speak- ing is an essential part of life for most conservation professionals. Indeed, as conservation action is increasingly based on multidisciplinary teams and stake- holder engagement, effective public-speaking skills are becoming ever more important. Here we will focus on the format for a typical conference presen- tation, but in broad brush strokes much of this is also applicable to classroom teaching, presentations to a local nature club, leading field trips, testifying at a government hearing, and other forums.[7.4] Conservation conference presenta- tions almost always have two elements – what you say and what you show for visual aids – and success depends on smoothly integrating the two.

Like a written paper, your presentation will probably follow the standard format of Introduction, Methods, Results, and Discussion, details of which we will cover in the next chapter on writing a paper. The main challenge is how

Using clear, concise language and aiming for a focused message will increase the amount of information that your audience actually understands and retains.

to sharply condense what you would cover in a written paper. You will have the most focused attention of your audience during the first 30 seconds, so use this time to capture them with a compelling story, statistic, analogy, or key outcome that piques their interest. Talks are typically allocated 15, 20, or 30 minutes, but you are usually expected to leave 3 or more minutes at the end for questions from the audience, so you may only have 12 minutes, or 25 at most. Skimming off the cream of what would appear in a written paper may take multiple attempts, each one getting shorter and shorter, but it is very important to get the timing right because session moderators are usually very strict about timing, especially at conferences where simultaneous sessions need to be kept in synchrony. It is not just a matter of cutting out details and thereby conveying less information. Indeed, by using clear, concise language and aiming for a focused message you will increase the amount of information that your audience actually understands and retains.

The worst case of poor timing I ever saw was at a major event, an international congress that met on a 4-year cycle, where a student had just finished her Introduction and started describing her Methods when she received a 2-minute warning. As she left the podium in tears I turned to a colleague seated next to me and said, "What a shame. I blame her advisor; he should have insisted on reviewing a dry run." My companion turned in his seat to address the person sitting behind me and said, "What do you think? Should you have had her practice?" I turned, and yes, her advisor was sitting behind me looking very morose but not very apologetic. "Well it was a good learning experience for her," he replied. I bit my tongue.

Pain does have a way of reinforcing some learning experiences, but you do not want to learn how to give talks through pain and one key is to rehearse the talk many times, initially on your own, and ultimately a couple of times in front of your advisor and fellow students. This is important both to get the timing right and to be sure to cover everything you want to say smoothly. Note that this is not the same thing as memorizing your talk. It might be a good idea

to memorize your opening line to get you going, but after that you will sound more natural if you just let the words flow. Similarly, in many disciplines it is considered poor form to read a talk, even though this is standard practice in some quarters, such as history and anthropology. One reason many presenters avoid reading their talks is that they typically have visual aids that function both to engage the audience and to subtly remind them what to say next.

Visual aids have evolved from a diversity of forms (blackboards, overhead transparencies, slides, props) to an overwhelming dominance of computer-generated and projected material widely known by the name of the leading software, Microsoft PowerPoint (though there are a number of good alternatives). Not too many years ago, visual aids at conservation conferences were dominated by photographs to depict study sites and systems, figures and tables to show results, and a few text slides, typically to present objectives and conclusions and to acknowledge funding agencies, field assistants, and so on. The advent of presentation software has opened up a kaleidoscope of possibilities to present text, figures, and photographs in a dazzling array of combinations. The problem is that many of these presentations have so much dazzle that it can be difficult to grasp the substance.

Most people have learned to avoid the most flamboyant, over-the-top features, like text that comes flying in from off-screen, but they still cannot resist certain temptations. Why show just one full-screen image of a representative study site when you can show a montage of all of them? Why have a simple text slide listing the three hypotheses you plan to test when you can surround them with thumbnail portraits of all your study species and have them appear in three different colors? The answer is simple, most of the time all that stuff becomes a distraction. It is much easier for your listener to absorb one or two ideas from each image and then move on to a new image. Note that we are not bashing technology *per se*; under some circumstances its capabilities can be extremely useful.

Last week I saw one of those cases where computer visualization technology was used really effectively. The speaker used animation of maps to show how winter bird distributions changed from year to year in response to weather patterns. It was a great example of where one good visual aid was worth a thousand words.

The problem lies with how people typically use the technology, but these issues can be easily minimized. In particular, if you limit the number of ideas per image it is much easier to keep your audience with you. For example, by first showing a photo of your study species while describing its natural history, then showing a separate graph of its population change, you eliminate the chance that some people will be studying the graph while you are still describing the species. In many cases, presentations actually force the audience to choose between listening carefully to the speaker or staying afloat in a tidal wave of visual material. This used to happen with older technology too, but it has become much more common now that you can generate an elaborate text image in a few seconds on your personal computer and decorate it with multiple images downloaded from the Internet. Keep in mind a basic principle of aesthetic design – simplicity is classy. Try drawing an outline of your talk with one box per per slide; write your key message for that slide, then think about the most compelling image to convey that message.

Having lots of text is particularly overwhelming, but as described above for posters, the visual impact of text can be greatly reduced if you use an outline format. For example, you can write "Territory mapping" or better yet, show a photo of a singing bird, rather than write "Bird populations were assessed biweekly using territory mapping of singing males." In the latter case you may end up reading the text along with your audience – a form of communication that works well for a parent reading to a child, but which has no place in a professional setting. It is rare that you need to display the exact wording of what you are saying; one exception might be presenting your hypotheses where close scrutiny of your precise language may be appropriate.

You are the most important part of your talk, not your visual aids.

In summary, presentation software is a fantastic technology if you do not misuse it. You will not go wrong if you focus on using the images as an ancillary aid to help your audience *listen to you* and *understand* what you are saying. Remember: *you* are the most important part of your talk, not your visual aids.

Four final thoughts about conference presentations

1 The most important thing to strive for is clarity. Ask yourself, would your grandparents be able to follow this presentation? Would they understand why your work is important for conservation?

2 It is also important to be engaging: look at your audience and talk to them directly. Tell them the story of your work. Even humor is okay in professional presentations if done in moderation.

3 Expect to be nervous. To some extent this will diminish with experience, but many very seasoned public speakers still get nervous. In fact, a totally relaxed speaker would probably not be very engaging.

4 Learn by observation, as illustrated by this anecdote:

> *When I first started making presentations as a graduate student, I couldn't figure out what I was supposed to do. All of the advice I was given by my advisory committee involved the content, not the presentation itself. My first few presentations did not go well, and then I realized that the best advice on how to give a good presentation was actually all around me. At that point I started watching and listening to other presentations with an eye and ear for what worked and what didn't. As a member of the audience, you are in the best possible position to see good and bad techniques in action and take notes for your future use.*

This approach covers the whole range of issues from the minutiae of what fonts and color schemes work well, to how you field difficult questions with grace and brevity.

Student paper competitions

Conferences sponsored by professional societies often have a competition to recognize the best student papers. If you have a solid research project that is nearing completion, you should seriously consider participating. If you are still in the middle of your program, attend a student paper contest to assess how to be competitive when you are ready. If that is not possible, with a bit of digging you should be able to find the names of previous competitors on the society's website and then find and read their abstracts. Be sure to follow the application instructions extremely carefully. When confronted with a large number of candidates, judges often start the filtering process by eliminating all the applications that did not follow the rules. The immediate payback for winning one of these prizes, perhaps a modest cash award or some books, may seem rather limited, but it is a very nice thing to add to your résumé and the reward may come in terms of a job offer.

Other opportunities for public speaking

If you really want to excel at giving a conference talk, practicing in front of a mirror or your advisor and friends is a mediocre substitute for speaking to a real audience. Elsewhere we have mentioned some opportunities that you should seize upon, such as presenting your project proposal and results in departmental seminars and serving as a teaching assistant. You also could volunteer to give a class guest lecture even if you do not have regular duties as a teaching assistant. Local nature clubs and civic organizations are often looking for guest speakers and there are probably some faculty members who could pave the way to an invitation if they knew you were interested in speaking opportunities. Did you go on an interesting trip the summer before starting grad school? Consider giving a casual presentation over lunch. Is there a public agency taking public comments at a nearby hearing? Go along and share your opinion. It does not take long to prepare a statement (they are often limited to 3–5 minutes) and seeing the public policy process unfold is a good experience even if you do not speak. The bottom line is that public speaking is part of most professional lives and with practice you will improve. Furthermore, as we will see in Chapter 10, it is one of the key vehicles for having an impact on conservation.

If you are one of the many people who are really unnerved by public speaking, this is all the more reason you should seek experience because you will become less nervous over time. You will learn how to read an audience, how to anticipate questions, and other skills that will help you for years to come. For most people, regular experience is enough to make them capable, if not excellent, speakers. In many cases, simply preparing thoroughly weeks in advance – especially, crafting a simple, clear message and honing your visual aids as described above – will go a long way in giving you confidence. If you are starting very low on the ladder or aspire to true excellence you may wish to seek special training, as did a student we knew:

> One of our department's new doctoral students gave a seminar on his master's research that was truly awful. He read from a prepared script at a pace that left everyone clueless about his research. Within the month his advisor had him signed up for regular sessions with Toastmasters International and a few years later he did a superb job when he presented his doctoral work.

Further readings

7.1 At this time, there are student conservation science conferences in Australia, China, Hungary, India, the US, and the UK. See www.sccs-cam.org for the original one at the University of Cambridge, UK, and links to the others.

7.2 There are hundreds of professional societies that are relevant to the conservation profession, making it impossible to list them all here. Your advisor will know the most relevant ones, but you can also search for them on the Internet if you use "society" or "association" as keywords along with a few choices to describe your interests. Some of the hits will be for environmental NGOs rather than professional societies so you will need to do some filtering. Relatively few professional societies are international in scope (notable exceptions include the Society for Conservation Biology and the International Society for Ecological Economics) and thus you should probably search with an adjective for your country (e.g., British) or continent (e.g., Asian), or both (e.g., American).

7.3 Given space constraints we can only offer general guidelines here. For further ideas on poster design and presentation, see colinpurrington.com/tips/poster-design and for general guidance on making presentations, and particularly the visual display of information, see:

Carter, M. 2012. *Designing Science Presentations: A Visual Guide to Figures, Papers, Slides, Posters, and More.* Academic Press, Cleveland, O.H.

Duarte, N. 2011. *Slide:ology, The Art and Science of Creating Great Presentations.* O'Reilly Media, Sebastopol, C.A.

7.4 Much of the material listed under note 7.3 is also applicable to oral presentations, and particularly the visual aids you will need. Also see:

Alley, M. 2013. *The Craft of Scientific Presentations: Critical Steps to Succeed and Critical Errors to Avoid.* Spring Press, New York, N.Y.

8 Writing papers

There are many ways to communicate important conservation messages –
television, radio, social media (e.g., Blog posts, Twitter, Facebook), lectures,
newspapers, magazines, books, and more – but professional papers in
peer-reviewed journals are fundamentally important. They are the foundation
of conservation science and also pave the way in other disciplines, such as law
and education. Professional articles also provide core information on which
other forms of communication, such as media interviews, are based.

To really understand what goes into a professional paper, it is critical to experience writing them, and thus this is an essential part of career development for
conservation professionals, including those who are not doing a research-based
degree. This is true even if your future job emphasizes reading journal papers
and your writing responsibilities shift to other outlets, such as internal reports
or creating websites. This chapter also goes beyond the process of writing scientific papers to include tactics for submitting an article to a journal, and coping
with peer review and rejection of your work. We begin with the most traditional
form of writing for graduate students, a thesis, then turn to writing professional
papers, a skill relevant to students (in both thesis and non-thesis programs) and
all conservation professionals

Saving the Earth as a Career: Advice on Becoming a Conservation Professional, Second Edition.
Malcolm L. Hunter, Jr., David B. Lindenmayer, and Aram J. K. Calhoun.
© 2016 John Wiley & Sons, Ltd. Published 2016 by John Wiley & Sons, Ltd.

A THESIS VERSUS A COLLECTION OF PAPERS

A thesis is almost always the basis for gaining a doctorate degree, is commonly required for a master's degree, and is occasionally required for a bachelor's degree. But writing and defending a thesis is not the end of the road. Most people strongly believe that science is not really done until it is published, particularly because, in this era of the

If you do not publish your work, you will significantly limit your contribution to science and conservation.

information super-highway, virtually no one reads theses after they are bound and deposited in a library. Your ground-breaking conservation research will remain a well-kept secret, locked in a thesis, until it appears in a journal.

Traditionally, a thesis was written as a single coherent document. However, an increasing number of institutions, especially in Europe and Australia, are allowing theses to be a collection of independent papers that are ready for publication or already published in journals. In many cases, a thesis also will include an initial chapter that sets the stage – like the Introduction of a paper – and a final chapter that provides a synthetic overview of the material presented in the previous chapters containing data. There are many advantages to writing a thesis as a set of connected chapters that are submitted as papers to journals. These include the following.

- The process of writing papers can often start relatively early in your program, ensuring that you can more quickly obtain good feedback from your advisors. Feedback from independent journal referees also provides useful feedback on thesis research and how it might be improved.
- Writing papers for journals demands conciseness, and this exercise will help you realize that not every minute detail needs to go into a thesis. This will make the possibility of completion seem a more tangible prospect.
- At the end of a degree program, you may be sick of the sight of your thesis. Finding motivation to return to it afterward and convert the chapters to papers can be difficult, especially given the demands of a new job.
- Having papers published during a degree program helps establish a strong track record, increasing the chances of successfully landing a job. Even if a career in academic conservation science is not the right one for you, demonstrating an ability to write clear, concise papers for professional publications will be a major advantage in almost any workplace.
- Finally, if a thesis comprises a series of published or in-press papers from peer-reviewed journals, it would be very difficult for thesis examiners to reject your work as unsound.

If you write your thesis as a series of papers, then it is important to be well versed in the skills of writing and submitting journal articles. These topics dominate the rest of this chapter because it is useful to have particular journals in mind when you are writing thesis chapters to ease the transition to journal papers.

Thesis appendices

If you write your thesis as a series of journal papers, you may need to include other detailed material in a series of appendices. Such appendices might include, for example, data on the abundance of each study species at each replicate study site, or the coordinates of the location for each of your survey plots. These data can be invaluable for people who want to extend your work in the near future (e.g., a new student working in the same study area) or perhaps replicate it many years in the future to understand how things have changed through time. Sometimes thesis examiners wish to check primary data too. Some journals have online supplementary sections or electronic archives where this material can be made readily available for readers even though it does not appear in the published paper. There is even a rapidly increasing number of journals that specialize in publishing data sets so that they are readily available for other researchers to view and re-analyze.

WRITING A PROFESSIONAL PAPER

The first essential step in writing anything is to envision your audience. Obviously, the readership of a professional journal is entirely different from the readership of a newspaper or popular magazine. Similarly, you will need to think about the readership of a particular journal because different journals have different audiences; we will return to this issue later in the section on selecting a journal. The next step is to write down the key messages you want to convey to avoid having them overwhelmed in all of the detail that will follow. Complicated writing usually has sections with subheadings to make the structure clear; for example, many scientific articles include an Abstract, Introduction, Methods, Results, Discussion, Acknowledgments, and Literature Cited. Other types of professional papers, like law review articles or articles in the social sciences, will also have a standardized format.

The key sections of a scientific paper

The **Title** of an article is critical; ask any publisher of novels how this makes or breaks a book. For example, compare the following.

An experimental investigation of the effects of temperature variation in turtle nests
High nest temperatures lead to more male hatchling turtles
Boys like it hot: nest temperature affects sex of turtle hatchlings

The first version uses the most words yet only conveys the topic of the paper. The second version provides both the topic and the key result. The third version gives the key result and may capture more interest with the opening part, although many editors would balk at the cute aspect of the title. With literally tens of thousands of articles published every year there is a lot of competition for the limited time that most people have for reading. Papers with an interesting and informative title will be those most likely to be read. Moreover, recent research shows that scientific articles with short titles are more likely to be cited by other scientists than those with longer titles.

The **Abstract** is by far the most important part of any scientific article. Only a minority of readers will read beyond the Abstract, so this must contain the key messages you want to convey in your paper. Usually this is done by summarizing each of the major parts of the paper – Introduction, Methods, Results, and Discussion – in one to four sentences each, essentially a micro-version of the paper. Choose

The Abstract is by far the most important part of any scientific article.

your words carefully because electronic searches of journal articles are usually based on words in the title and abstract. These search engines often provide the Abstract of a paper to readers, another reason why it must present your critical information. For conference proceedings, the Abstract will probably be the only written version of your presentation (see Chapter 7). Almost all journals have a strict word limit for the Abstract, for example 250 words, and within this limit it will be a real challenge to clearly describe your study, emphasizing the new findings. Because it is so important, many authors write the Abstract last to be sure it captures the key results and conclusions of the paper.

The **Keywords** section usually asks for 5–10 key words or phrases that characterize your article. This is an increasingly important part of the paper because it provides the basis on which other people can conduct an electronic search for a given topic. The selection of the "right" key words increases the chance that your work will be found amongst the avalanche of other published articles in "journal land." Because most electronic searches also include the title, try to maximize the reach of your article by not duplicating keywords with those in the title.

The **Introduction** must explain to the reader why your piece of work is important to conservation, or they may stop reading. This is usually done by describing the bigger picture and then focusing down to what you actually did. Envision an inverted pyramid, ▼; start broad to set up the paper and end with the key questions or hypotheses you have posed (see Chapter 6 for an example in the context of proposal writing). This structure should make it clear how your work targets gaps in knowledge or understanding that have emerged from other work. Aim to highlight what is different about your new study, and also how it builds on the body of previous work. Although it is useful to summarize previous research in the Introduction of your paper, this should not be exhaustive. Brevity is a key part of successful scientific writing and taking too long to cut to the chase weakens the Introduction of many scientific papers.

The **Methods** section describes where the study took place and how the work was done and similar details. Subsections might include a description of the study area, the techniques used to gather data, and the computer-simulation model or statistical-analysis methods you employed. The Methods section should, in theory, provide sufficient detail to enable another researcher to replicate your work. However, given the shortage of journal space, most Methods sections only provide a general understanding of what was done and someone wishing to replicate the study would have to contact you for more details. Sometimes additional details about the methods can be provided in an appendix or supplemental information, often published online. If you are using a method that will be known to your readers, citing a key paper describing that method can save space. If the method is unique and takes more than a couple paragraphs to describe, it may be a good idea to write a separate paper describing the new method in detail.

The **Results** section outlines what you found. Make sure that your subject matter, not the statistics, tells the story. Statistics are important, but they should be used to support your case and story, not lead it. Compare the following two sentences.

Nested ANOVA demonstrated that fish density in Pond A was significantly greater than in Pond B ($P < 0.05$).

The density of fish in Pond A , $5.7/m^2 \pm 0.7$, was significantly greater than in Pond B, $3.7/m^2 \pm 0.3$ (ANOVA, $P < 0.05$).

In the first case, statistics drive the sentence and a key biological attribute, the estimated density of fish populations, is not even reported. In the second case, the biology drove the sentence, with confirming support from the statistics.

Often you will have many findings. Try to set these out in some kind of logical sequence, perhaps with sub-headings if each section is longer than a paragraph or two.

> General findings – numbers of species and individuals recorded
> Yellow-footed ankle-biter habitat
> Red-bellied ankle-biter habitat
> Purple-shinned ankle-biter habitat
> Ankle-biter community structure

Simple statements of your findings are useful, but it is also important for you to emphasize the key new results of your work. Do not expect readers to discern a pattern from a blizzard of numbers for themselves; you need to make the key trends and patterns apparent to them. Given the shortage of journal space, your Results section needs to be data-rich without redundancy. For example, this will force you to choose between presenting the same information in a table or a figure. Figures are often preferable for displaying patterns, as per the old saying "one picture is worth a thousand words," but one can often fit more information into a table. Most journals now limit the number of figures and tables that you can include in a paper, as well as overall paper length.

In some cases, it may be necessary for you to present a large number of values, for example the actual densities of multiple species at multiple sites at multiple times or the full results of a social survey with dozens of questions. You may be able to put these in an Appendix or the journal's electronic archive, or they may have to remain in a thesis appendix.

Finally, note that the Results section only presents your findings. Many novice writers make the mistake of interpreting these findings prematurely, but this should usually be reserved for the Discussion. (We note that a combined Results and Discussion section is sometimes acceptable, especially in the social sciences.)

The **Discussion** section is where you interpret your results by summarizing your main findings and describing how they relate to the questions that were posed. This section should not repeat all your Results, but rather present them synthetically in a larger context. For example, you might want to begin by directly answering the questions (or accepting/rejecting the hypotheses) you set out at the end of the Introduction. The Discussion also may outline how your findings differ from (or concur with or complement) those of others. For conservation journals, the Discussion is also an important place for you to explain the implications of the results for conservation.

A common problem in many papers is that the Discussion section is far too long. It is often the section that can be edited down 20–30% in length without any loss of useful content of a paper. A journal editor will almost always ask you to stick closely to the key implications of your work rather than speculating widely about all kinds of topics that are only loosely linked to the actual data you obtained.

The **Acknowledgments** section is best kept short, three to four sentences if possible. Here you need to thank your project funders, and anyone who made a significant contribution to your work and who is not a co-author, such as research assistants and manuscript reviewers. Many journals also require you to provide details of ethics approvals in the Acknowledgments. While ensuring that acknowledgements are short, also take care not to leave out key people. Of course, you need to have some judgment here; for example, it is not essential to write: " … and finally, last but not least, thank you to my parents for making me possible."

The **Literature cited** will include all the papers you have referenced in your article. Make sure that the format matches absolutely precisely that required by the journal. Unfortunately, it seems like almost every journal does citations a bit differently, a form of "cultural diversity." Computer packages that store databases of papers and linked details like Endnote® can make this a relatively painless task. Note that some journals will even request that you cite other papers they have published, a rather dubious practice designed to increase a journal's impact factor.

The ways different people write a paper

Different people have different recipes for the way they write. Some choose to outline each section of a paper and then go back and fill out the text. Others write each section sequentially in fully written form, sometimes handwritten, and then input to a computer later. Yet others start with the easiest section first (often the Methods) and then tackle the most difficult ones last, or vice versa. There are no generic recipes that work for all authors, but virtually everyone requires many iterations of a manuscript to bring it to a stage where it is sufficiently polished to submit to a journal. Thus you need not strive for perfection on the first draft.

Once you have written a complete draft of your paper, go back and read it thoroughly and try to shorten it by at least 10–20%. This is also the time for polishing your prose, using some of the ideas presented in Box 8.1. Remember, brevity is a pivotal part of effective science communication. Many journals set a maximum word limit, but it is wise to aim well below this if you can. Ask yourself, what is new and different about the article? How does it contribute to conservation knowledge? Does your article clearly communicate the new discoveries you have made?

Box 8.1 The art of writing

Good writing is not easy. It can seem tough to convey what are often complex messages with simple language and grammar. It is often surprising how many students have enormous difficulties with technical writing and advisors are not immune from writing problems either (including ourselves on many occasions).

Many books[8.1] have been written on how to write and it is well beyond the scope of this book to repeat what is in them. Here we simply present some basics to reinforce what should be straightforward for many students.

1 **Write short sentences; try to make the average sentence about 15–20 words.**

For example, the mean sentence length of this chapter is ~18 words and for others in this book it is about 18–23 words.

2 **Use plain language that your readers are likely to understand.**

For example, the terms habitat specialist and habitat generalist would usually be preferable to stenotypic and eurytypic for most conservation-science writing.

3 **Remove excess or useless words.**

Compare

> We tackled a series of key questions in order to attempt to resolve major issues associated with the melting of chocolate frogs.

(continued)

Box 8.1 *The art of writing (continued)*

with

> We addressed three key questions on melting chocolate frogs.

4 Use an active voice and try to make your key points positively rather than negatively.

As an example: "We documented X ... " rather than "The results were not adequate to document Y ... ".

5 Use vertical lists to break up complicated text.

We posed three key questions.

   (i) Do chocolate frogs melt at high temperatures?

   (ii) Do melted frogs solidify following initial changes in their state?

   (iii) Will temperature changes alter the conservation status of chocolate frogs?

I often find I get too close to a paper after weeks of trying to perfect it. If time allows I like to put it away for a while and forget about it. Then I will pull it out and read it aloud slowly to myself. It's amazing how many errors surface when I do that. It is after that stage that I feel that I can give it to my colleagues to get critical feedback from them.

Of course, it is essential to ask others to read your manuscript and critique it after you are reasonably content with it. If they do not understand certain points, assume that you were not clear, not that they are stupid! Authors often become so close to their papers they are unable to be self-critical. Try not to be offended by criticism of your masterpiece; remember that critique and robust peer review is integral to the scientific process and better writing. It is preferable to receive criticism first from people who are close to you than to hear it later from thesis examiners or journal referees.

I have to chuckle thinking about it now, but the first time I got comments on a chapter from my advisor, I was absolutely shell-shocked. I took them home and stamped around the living room jumping on the paper. The crumpled pages stayed on the floor for a day – my housemates thought I had totally lost it. Of course, the comments were spot on and a few days later I began working in the suggested edits.

AUTHORSHIP

The issue of authorship can create significant problems between students and their advisor. Single authorship is common in the humanities, but relatively few papers in conservation science these days have a single author. Perhaps this reflects the fact that the combined wisdom of multiple minds is often needed to pull together a good article that is worth publishing in a journal. Different fields have different recipes for assigning authorship. In economics and mathematics it is usually done alphabetically. In the biological sciences, authors are usually listed in order of the extent of their contribution. A student should almost always be the first author on an article based on their thesis work as they have done the bulk of the work. One exception might occur if a student has disappeared after graduation, leaving the advisor a substantial task in preparing thesis chapters for publication. Other authors should only be listed on a paper if they have made a substantial contribution. Of course, much of science is about ideas, so a significant contribution might be the underpinning ideas that instigated your thesis research. Strategically, it is important to realize that publishing is the life-blood of many academics' reason for being, and they will probably offer you more thorough help if there is the prospect of co-authorship of articles. Many relationships between students and advisors have suffered serious damage because open conversations about authorship

Many relationships between students and advisors have suffered serious damage because open conversations about authorship were not initiated.

were not initiated. Thus, if your advisor does not raise the issue early in your academic program, you should.

SELECTING A JOURNAL FOR YOUR PAPER

Who reads your target journal? Who are you trying to reach with your information? Very early in the process of drafting a paper you need to think about the journal you are targeting because this will shape the emphasis of your paper.

I was writing a paper on amphibian populations in managed forests and I considered Conservation Biology, Journal of Wildlife Management, Journal of Herpetology, *and* Forest Ecology and Management *as possible places to send it. Each has quite a different readership and I quickly realized that the Introduction and Discussion sections would have to be written and tailored quite differently depending on which journal I finally decided on. For example, the two opening sentences below would be better suited to* Conservation Biology *and* Forest Ecology and Management, *respectively:*

Maintenance of biodiversity cannot be exclusively focused on reserves; it must be integrated with commercial activities in unprotected areas such as logging, farming, and fishing.

Wise forest management must sustain both timber production and other values such as biodiversity conservation, including amphibians and other taxa that are often overlooked.

Choosing the right journal is a key issue. Large numbers of papers are rejected (and often not even sent to referees by the Editor) simply because the paper was submitted to the wrong journal. Some differences are obvious when you read a journal's guidelines to authors. Some will become obvious only from talking to

Large numbers of papers are rejected simply because the paper was submitted to the wrong journal.

experienced authors and carefully studying the types of papers published by a journal. If, after you have done some homework, you are still unsure about whether your paper is suitable for a particular journal, then e-mail an abstract of your paper to the Editor or a member of the Editorial Board and ask for their advice. Beyond the appropriate subject matter, there are some other issues to consider.

Impact factor

A journal's stature or prominence should probably be a significant consideration in your selection process. The most widely used measure of this is the impact factor calculated by Science Citation Index (SCI), which reflects the average number of times that papers in a given journal are cited. Impact factors clearly demonstrate that an article in *Science* or *Nature* will be more widely cited (and presumably read) than a paper in a narrower journal such as *Environmental and Resource Economics* or *Journal of Mammalogy*. Good papers in lower-impact journals can still be highly cited and poor papers in high-impact journals can go unnoticed or be heavily criticized. However, impact factors are far from the be all and end all of science publishing. Indeed, some people are increasingly critical of bibliometric assessments of "research value" and suggest that they can lead to a reduction in the value of science.[8.2]

Many authors target high-impact journals and hence these journals have a concomitantly high rejection rate, often over 75%, sometimes over 95%. Thus, while it is good to aim high, it is unwise to aim too high and you should stay grounded in reality. With many journals taking 2–6 months to review a manuscript you could be wasting significant time (both yours and the editor's and referees') pursuing a high-impact journal and would be better off aiming lower. Moreover, lower-impact journals still have a very valuable role to play in conservation.

Geography and scope of relevance

Geography is another important factor to consider when selecting a journal. You can start by asking yourself if your results transcend your particular focus and will be of interest to a global audience. For example, can you make a really convincing case in your Introduction and Discussion that your research on the attitudes of local people toward national parks in Namibia will be of interest to park managers in Nepal and Nicaragua? If so, then international journals such as *Conservation Biology, Conservation Letters,* or *Ambio* may well be appropriate. In other cases, your work might be more relevant to workers in a particular region for which a regionally focused journal such as *Pacific Conservation Biology* will be the best outlet. Note that some regional journals, like the *Canadian Journal of Economics* or *Australian Journal of Zoology*, accept papers from around the world but are more likely to be read by Canadians and Australians, respectively. These journals can still be accessed widely due to electronic indices. In many cases, a journal oriented toward a particular taxon, type of ecosystem, or discipline (e.g., birds, forests, or sociology, respectively) will provide your best audience. Most of these will accept a conservation-oriented paper as long as the science is sound and innovative.

Turnaround time

The process of submitting a paper, getting referees' reports, re-submitting the modified manuscript, being notified of acceptance, and then finally seeing your masterpiece in print can be a prolonged one – sometimes seemingly glacial in rate of progress! Some journals have much faster submission to publication rates than others; they range from a few months to years, with a year or two being typical. Thus, it is worth the effort to calculate turnaround times by examining the submission, acceptance, and publication dates for a sample of papers from each journal you are considering. Note that a significant portion of turnaround time rests with authors who take months to revise a paper after initial review, so this is one part of the process that you can control.

Circulation

All other things being equal, it is preferable to have your paper appear in a widely read journal. Open Access journals (e.g., PLOS One) allow a paper to be readily found by anyone with web access, notably colleagues from developing countries and many mainstream and science journalists. A number of journals allow authors to buy Open Access status for papers, but this can be expensive.

The number of libraries that subscribe to a given journal can be determined through search engines like WorldCat.[8.3] You might also want to favor journals sponsored by a professional society rather than a purely commercial publisher, because these tend to have large numbers of personal subscribers who are members of the society.

SUBMITTING A PAPER TO A JOURNAL

Once you have selected a journal, there are some important things to consider, starting with formatting. Virtually every journal has its own idiosyncratic format. If you have not followed it, the referees may wonder how carefully the rest of the paper has been written. Some journals will not send out a paper for review if the formatting is not correct. Of course, clear writing is essential and, as we have stressed above, so is brevity. All journals have strict length limits and journal editors prefer papers that are well under these limits.

Another key consideration is the cover letter that will accompany your paper. This should succinctly explain what is new and exciting about the work and how the paper is well suited to the journal you have chosen. All covering letters require a statement saying that the work has not previously been published and is not being considered for publication elsewhere. The cover letter might suggest suitable referees who have not already seen the paper and provide their contact information. Many editors welcome this because it is increasingly difficult to find referees for articles, although others are now wary because of the potential for referee fraud. The cover letter should be well written and mistake free, as it is the first thing the editor will read.

YOUR PAPER COMES BACK FROM THE JOURNAL

Part 1: coping with rejection

Being a scientist requires dealing with rejection. Some of your ideas get rejected, some of your grant applications get rejected, and – particularly – your papers get rejected. There is nothing unusual about having scientific papers rejected; in fact it's the norm. All conservation scientists (including many very famous and influential ones) have had some of their papers "bounced" by journals. Leading scientists such as the late Robert Macarthur and the late Graeme Caughley both believed that if at least some of their papers were not rejected, then they either were not publishing enough or not pushing the boundaries of the field of ecology far enough.

I heard recently that the first paper describing the Krebs cycle was rejected by Nature. *Another example was that of a close colleague of mine who wrote a cutting-edge paper on acid sulfate soils.* Nature *rejected it, but then later reported it as a brilliant and important new finding in their News section after the article appeared in another journal.*

Pat: "Can we have a chat about the chocolate frogs paper we submitted to the *Journal of Confectionary Conservation*?"

Terry: "Sure, I have some time now."

Pat: "The journal bounced it. I just can't believe the comments from the referees and the editor! They completely misunderstood what we were saying!!"

Terry: "OK, sounds pretty unfair, but don't take the rejection thing personally. Everyone has their papers bounced – even the gurus. And sometimes really good ones that push the boundaries of scientific thinking have to do the rounds of several journals."

Pat: "So, we've got comments from two referees. The first one said it was a great paper; just that the Discussion was too long and the Methods and Results needed to be better expressed. The second reviewer absolutely belted the paper and made all kinds of nasty comments on the way we collected the data and the techniques for analysis. The Editor has clearly sided with the second referee."

Terry: "They often do. Did she say anything else?"

Pat: "Well, she did say that the paper had some merit and she recommended we try a couple of other journals. But I still just can't believe this ... the amount of work that went into this manuscript and then for it to get bounced like that!"

Terry: "Hey, don't take it personally. It is a terrific paper and we aimed high; you know the *Journal of Confectionary Conservation* has a rejection rate of about 85%. Did she mention the *Annals of Conservation of Confectionary?*"

Pat: "Yep – *Annals* was mentioned, but writing a new manuscript will take forever; it took me five solid months to write this paper."

Terry: "You don't have to start afresh. Just use the comments to improve the manuscript, and change the formatting for *Annals*."

Pat: "What about the Discussion?"

Terry: "Well it's probably true that the Discussion section is too long. I'd get rid of those paragraphs about how chocolate frogs can melt before you eat them on hot days and where the wrappers could go. I know it seems important, but most readers consider it a separate issue."

Pat: "What about the randomized block design for comparing melting rates of jellied and chocolate frogs? The second referee was super-critical about that."

Terry: "Yes, it's possible that Referee 2 was lazy, but it's better to assume that the fault was ours for not being clear enough. So, let's make it clearer what we did and how that approach for comparative testing is novel."

Pat: "Hmmm. OK. Perhaps another couple week's work would be enough for re-submission."

Terry: "That's not much compared to the years you have worked on this."

Understanding the refereeing process

Nobody likes rejection. Nevertheless, understanding and dealing with rejection will be easier with some understanding of the journal editing and refereeing process. As outlined above, many of the leading journals have massive rejection rates. Take a moment to think about the life of an editor of a journal. It is no fun rejecting other people's work, especially when they have invested months and often years in it. Many authors have large and surprisingly brittle egos, and they are often quick to get grumpy or even deeply offended. However, the reality is that journal space is limited and there is rapidly increasing competition for that limited space.

Although the peer-review process is a human process and inherently imperfect, it is still the best one we presently have in science publishing. The sheer number of papers being submitted and the corresponding pressure on editors and referees may be increasing the incidence of Type I and Type II errors. That is, the number of good papers that are rejected and the number of bad papers that are accepted. This is unlikely to improve as long as the system depends largely on overworked, underpaid editors plus sub-editors and referees who are not paid at all. You may have the opportunity to be a referee soon after you publish your first paper; it will be a good experience to have under your belt.

Referees are often asked to assess many papers and it is increasingly difficult for them to examine them in depth, a process that can take days of work to do well. Thus, sometimes referees do not read papers as carefully as they might. On the other hand, it is probably more common that authors do not communicate their ideas as clearly as they should, especially given that the average reader will probably be less careful than the average referee.

Two inter-related character traits are critical in dealing with rejection – resilience and persistence. Resilience is essential in conservation science and it is part of bouncing back from rejection and refusing to become disillusioned. The other trait, persistence, is equally important, as illustrated in the dialogue between Pat and Terry. If you are convinced your ideas are correct

Persistence and resilience are keys to getting your work published.

and important, then stick with them. If you are rejected by a journal, then work with the referees' comments and input from your colleagues to improve the paper and submit it elsewhere. If your work was sound, then you will eventually have success somewhere. The impact factor for the journal may not be as high as the one you chose initially, but others will notice your paper, especially in this era of search engines, and particularly if you choose a good title, select the right keywords, and write a compelling abstract. Of course, some papers are discovered to have a fatal flaw and cannot be published. This is unfortunate but not an utter disaster; most scientists have a few papers in their files and are glad they died a quiet death.

Part 2: revising after possible acceptance

Many journals, particularly the high-impact ones, virtually never accept a paper on first submission. The letter returning your paper is likely to contain some variation of the following language.

> We are unable to accept your paper in its present form. However, we invite you to carefully consider the thought-provoking comments of the three reviewers. Should you be willing to substantially revise your manuscript, then please ensure that it is returned to us within 4 months for reconsideration for publication; otherwise it will be treated as a new submission. If you choose to re-submit then please describe in detail how you have addressed the comments of the referees.

This can be re-interpreted as: "We will reconsider your paper, but you need to do a lot of work on it." Of course, this needs to be treated seriously because an invitation to revise and re-submit is by no means a guarantee of ultimate acceptance. Two critical tasks are needed to reduce the chances of being rejected twice: (1) actually completing a comprehensive revision of your manuscript; and (2) documenting in detail the changes you have made as part of your re-submission. One way to facilitate this process is to construct a table like Table 8.1 to include with your cover letter. If you disagree with a suggestion, do not ignore it. Say that you have considered this suggestion but decided not to follow it, and explain why explicitly.

Be aware that many papers (up to 30% in some journals) are rejected outright when they have been re-submitted, even if you have addressed all of the comments of the referees.

Table 8.1 Responding to reviewers' comments

Comment	Response
Reviewer #1	
Typographical errors	Five typographical and grammatical errors were identified by referee #1. These have been addressed.
Reduce the length of the Abstract	The Abstract length has been shortened by 45 words.
Reduce the emphasis on fire	Over half of the material on fire has been removed.
Remove the Conclusion section	This section has been deleted.
Check the citations	Two missing citations from the text have been added to the Reference list.
Delete Table 1	This is the only suggestion with which we disagreed. We believe that many workers will want to see the Checklist in tabular form as a quick "look-up" guide. This is certainly the response we have had with five practitioners we consulted.
Referee #2	
Expand the basis for the paper with greater reference to material around the world	The over-focus of the original manuscript on forests from North America and Australia has been remedied. Studies from seven other countries have been cited.

I recall a recent horrendous experience where we received very favorable comments on a paper that was sent to a leading journal. The referees provided some very constructive comments that led us to spend 4 months re-analyzing the data sets to improve the paper. When the manuscript was re-submitted it was sent out for re-review to a new set of referees who hated it. Needless to say, after the paper appearing to be close to being accepted the first time around, we were profoundly disappointed that it was resoundingly rejected during the second review, but the work is scientifically and statistically sound and we will persist in the hope that it will one day find a good home.

Part 3: when your paper is finally accepted

Getting a paper published in a journal can be a long and arduous process. It typically will involve at least one significant revision, and perhaps submission to multiple journals. Therefore, final acceptance should be a cause for celebration; take time out with some colleagues to congratulate yourself. Most journals will provide you with an electronic version of your paper, such as a PDF, that you can send to others in your field and especially your parents. Spreading the word yourself can be important because conservation professionals are finding it increasingly hard to keep up with the stream of new publications. Having paper reprints can also be useful for people to take away and read after you give a seminar or present a poster at a conference. More people will get to know about your work that way and this might lead to other opportunities for jobs or research.

OTHER KINDS OF PUBLICATIONS

This chapter has focused predominantly on one kind of publishing – in scientific journals – and many of the general principles are also applicable to other professional journals, such as in law or education. Although professional writing is critical for developing your career as a conservation professional, most journal papers are read by few people other than your colleagues.

It is essential to be cognizant of the audience and who you are trying to influence with your writing.

For example, despite the best hopes of the conservation community, policy-makers only very rarely (if ever) read journals such as *Conservation Letters* and *Ecological Economics*. Therefore, forms of written communication other than scientific articles, such as popular books, articles in magazines and newspapers, blogs, and brochures, play a critical role in conservation. Box 8.2 provides an overview of the social media realm. They represent an opportunity for you to make a contribution, as we shall see in Chapter 10. Many of the approaches to better writing that were discussed earlier in this chapter can be important in other kinds of written communication, although the writing style and presentation will vary depending on the audience.[8.4] As always, it is essential to be cognizant of the audience and who you are trying to influence with your writing.

Box 8.2 *Social media and Saving the Earth by Lachlan McBurney and Thea O'Loughlin*

The "field" of social media includes blogging, Twitter, Facebook, and online publications like The Conversation. It has many benefits and huge reach, but also lots of traps. This is a rapidly evolving space through which even the most experienced "users" are still feeling their way.[8.5] There are some important advantages of social media. These include:

- Sharing new publication links, interesting articles, pictures from the field, and asking specialist questions.

- Networking: there are many success stories of problems solved by social media: for example modelers seeking data sets, researchers seeking volunteers, discussion of experimental design, projects seeking partnerships, and species needing identification. Of course, there are also the social benefits of developing support networks with others working on saving the environment from around the world.

- Finding employment: social media is increasingly used in conjunction with standard job websites.

- Sharing early views of papers and reducing the typical 2–3 year wait for peak citations.

- Promoting the adoption of conservation in policy arenas.

- Quickly rebutting incorrect information.

- Reducing the risk of science stories being misrepresented by journalists.

- Drawing attention to underrepresented areas (e.g., women in science).

Like all media there are some important things to remember before generating social media.

- Make a good first impression. In the case of a blog, the first line/paragraph of is what is first seen (especially when also shared on twitter or other media). Conversely, don't start with a "loser" line such as "I haven't written a blog in a while so I thought I should … ". Always write in a way that does not alienate readers. Always include a great picture.

- Use careful editing to protect you from publishing an error (online is forever!) or a hastily considered piece. Rash posts may be regretted later.

(continued)

> ## Box 8.2 Social media and Saving the Earth by Lachlan McBurney and Thea O'Loughlin (continued)
>
> - Be aware that producing good content for social media and then administering it can be time consuming.
> - Only a subset of your audience will be reached with social media – readers are more likely to be young and urban.

Further readings and notes

8.1 There are many excellent guidelines for writing professional papers. Our five principles for plain writing (Box 8.1) are distilled from Cutts (2013).

Cutts, M. 2013. *Oxford Guide to Plain English*, 4th edn. Oxford University Press, Oxford.

Day, R.A. and B. Gastel. 2012. *How to Write and Publish a Scientific Paper*, 7th edn. Cambridge University Press, Cambridge.

Powell, K. 2010. Publish like a pro. *Nature*, **467**, 873–5.

Sand-Jensen, K. 2007. How to write consistently boring scientific literature. *Oikos*, **116**, 723–7.

8.2 Werner, R. 2015. The focus on bibliometrics makes papers less useful. *Nature*, **517**, 245.

8.3 Access to WorldCat may be limited, but your librarian can help you obtain information on the circulation numbers of journals. Many periodicals print this information in the last issue of each year.

8.4 Tailoring conservation messages to a popular audience is covered in Jacobson (2009) and Chapter 10 of Jacobson et al. (2006).

Jacobson, S.K. 2009. *Communication Skills for Conservation Professionals*, 2nd edn. Island Press, Washington D.C.

Jacobson, S.K., M.D. McDuff, and M.C. Monroe. 2006. *Conservation Education and Outreach Techniques*. Oxford University Press, New York, N.Y.

8.5 For further ideas about social media see: http://www.**nature**.com/news/online-collaboration-scientists-and-the-social-network-1.15711; http://www.sciencedirect.com/science/article/pii/S0962892414001342; http://jmq.sagepub.com/content/early/2014/09/12/1077699014550092.abstract; http://theconversation.com/a-better-formula-for-science-communication-222; and http://www.southernfried science.com/?p=18088 – a case study of how a relatively normal paper went viral. http://ianluntecology.com/ provides some thoughts by Ian Lunt who is a current guru in ecology and conservation blog writing.

9 Finding a job

If you have arrived at this chapter directly from the Contents list, that is okay; it is certainly no worse than skipping to the last page of a novel to see how it ends. Of course, eventually you are going to have to work through many, if not all, of the preparations described in the earlier chapters before you are ready to find a long-term job as a conservation professional. Indeed, that is the punch-line of this chapter. To be more explicit, the key to finding a good job is developing skills; the mechanics of the job search are secondary. This is not to say that the process of looking for a job does not require your close attention, but you should try to keep it in perspective.

WHAT TO SEEK

Hopefully, by the time you have finished a university degree you will have a pretty good idea of your job prospects across the broad realm of conservation. You will know where your topical niche lies within the breadth of conservation and you should have a feel for opportunities to pursue this. You can read about opportunities at websites like those described under Further readings and notes,[9.1] but you will have a much more robust understanding if you have discussed careers with conservation professionals from a variety of organizations. You may recall from Chapter 5 that one idea for investing in your department and yourself included organizing sessions to discuss careers, and the perspectives of people from outside of academia may be particularly valuable.

It is possible that you have arrived at this point in your life with all of your experience in academia. This might happen if, for example, you worked as a

university research assistant during all of your undergraduate summers. If this is true, and especially if you are on track to earn a doctorate degree, you may be suffering from tunnel vision with respect to your future. You may assume that the only reasonable goal is a faculty position. Academia is certainly part of the conservation enterprise, but it does not lie at the center by any means. Academics are like the support staff of conservation: chiefly they educate the next generation of conservation professionals and undertake research that hopefully informs conservation action. These are vital activities, but they are a step removed from the front lines of conservation. Some academics manage to become more directly involved through special roles, such as serving on a policy-advisory group for the government, but many do little more than you would expect of an activist citizen, such as writing letters to elected representatives.

What about the academic experiences that you may have accumulated as a graduate student, are they wasted if you leave academia?

If you want to focus your work on conservation look beyond academia.

Absolutely not. First, the organizational, management, and communication skills you have developed will serve you well in any type of professional employment. Second, because conservation must be informed by sound scholarship, your academic training will always be relevant. For example, if your graduate program has allowed you to experience first-hand how the process of scientific research unfolds, this will allow you to assess scientific studies more critically. Finally, there are significant opportunities to undertake research and education outside of academia.

Broadening your horizons to include government agencies, environmental groups, and consulting firms will increase the pool of jobs by at least ten-fold. You might be most familiar with prominent organizations, such as your national conservation agency or the World Wide Fund for Nature, but these are just the tip of the iceberg. Collectively thousands of smaller organizations hire more people than the few major groups and it is often easier to land your first long-term job with one of them. There are also a surprising number of conservation-related jobs available in organizations that may seem to lie outside the mainstream; for example, museums, ecotourism firms, charitable foundations, law firms, zoos, aquaria, botanical gardens, publishers, professional societies, and more. Furthermore, organizations based on natural resource extraction have long hired conservation professionals such as foresters and hydrologists to help ensure sustainable resource use. With ever-growing pressure to be responsible stewards, many of these groups are also hiring people specifically to help ameliorate their impact on the overall environment. For example, some timber companies hire biologists to work on integrating timber production and maintenance of biodiversity. Note that as you move further away from academia you will be evaluated more in terms of all those skills mentioned in Chapter 5 – leadership and communication and so on – and less on your specialized knowledge.

To restate what we said in Chapter 1, conservation is a large and diverse undertaking and with the proper skills you can find a suitable niche. It may be rather different from the academic environment that has been your home for many years, but you can find an opportunity to make a genuine contribution, especially if you search widely. Searching widely also means being open to unconventional approaches, as illustrated by the following.

> *One of my undergraduate advisees was tired of all things academic and looking for some new experiences after she finished her degree in Ecology and Environmental Sciences, so she took a job as a cook on a small ship. It turned out to be an oceanographic research ship taking hydrographic soundings to update nautical charts. Soon the ship's technical team leader learned of her training and urged her to get scuba certification so that he could assign her to work on the technical team. After 2 years of this she realized what her life's work would be. She moved to the shore of another ocean, earned a Master's degree in Coastal Zone Management, and now works for a state agency on coastal conservation issues.*

This anecdote also illustrates another important idea: the indirect approach. If you aspire to work for a particular organization, but your ideal position within that organization is not available, it may make sense to apply for a different position. There is a good chance that within a few years you can reach your goal through lateral transfer, relocation, or promotion. In other words, it is usually easier to move around within "the house" than to choose your seat at the table while you are still standing on the street. The most common example of this is the process by which temporary appointments often evolve into permanent positions.

Conservation is a large and diverse undertaking and with the proper skills you can find a suitable niche.

HOW TO SEARCH

You probably already know the most efficient ways to scan large numbers of job announcements: to a significant extent they are those same websites that you have used in the past to look for summer jobs and graduate-student openings. A complete review of these is beyond the scope of this book, especially given our global audience, but you will find an introduction to these resources in Further readings and notes.[9.1] In most parts of the world, fair employment laws dictate that government agencies and organizations that do business with the government (which is the large majority) must advertise jobs widely. There are often two exceptions to these laws: temporary jobs and jobs in small organizations or businesses are often not required to be advertised, but many are widely posted anyway. Large organizations often have a portion of their website devoted to listing their job openings, so these may be particularly appropriate if you are keen on one organization. Advertisements in printed media are also extensively used, such as in the journals *Science* and *Nature*, professional society newsletters, and local newspapers, but even these outlets are largely Internet-based now.

Although the public marketplace of jobs dominates the scene, there is a small, subterranean stream of jobs that are never publicly announced or are only announced after they have been effectively promised to someone. Finding out about these jobs will require personal contact of some sort, and thus professional networking becomes a key skill. You can begin by sending a short e-mail to all those professional contacts you have fostered: your advisory committee, people you have worked with in the past, your undergraduate advisor, other faculty members from your undergraduate institution, people you have met at conferences, and so on. Some of these might be in a position to employ you directly, but the main thing is to let the world know that you are available. Most professionals receive many forwarded e-mails announcing jobs and they may start forwarding appropriate ones to you if they know you are job-hunting. Your e-mails can be quite succinct: just let people know when you will be finishing your degree, your geographic constraints (if any), the types of work you are seeking, and ask them to inform you of any opportunities. Be sure to attach your résumé.

You can also try visiting organizations to ask people face-to-face about job prospects. You will encounter lots of "Sorry, we have nothing open now ... " but you might get lucky, you will learn something about the organization, you could hear about leads at other organizations, and if you make a good impression it will help when a job does become available. If you have a gap between finishing your degree and finding a long-term job, you could take a temporary assignment with one of your target organizations or even volunteer; these experiences regularly morph into regular employment for people who demonstrate their talent and commitment.

The issue of geography returns us to an issue that was covered in Chapter 1: where you want to work and live. Conservation jobs are well distributed across the globe, but rather thinly in most places. Therefore, as a newcomer to the job scene, you may have to be quite flexible about where you are willing to work. For example, your heart may be in the mountains but it may be necessary to spend your first few years in the capital city developing your credentials before one of those uncommon mountain jobs comes along.

WHEN TO APPLY

In the best of all possible worlds your dream job will await you a few weeks after you finish your degree, just long enough for a relaxing trip to some wild place. Unfortunately, the timing will probably be less than ideal. In particular, many graduate students start to worry about a job months or even years too early.

Pat: "Could you write a letter of reference for a job, working up north with the Department of Conservation?"

Terry: "Sure, but what's the timing?"

Pat: "They're going to start reviewing applications next week, so if you could e-mail your letter by next Monday that should be fine."

Terry: "Next week?! If you get this job they'll probably want you to start within a month and you haven't even finished analyzing your data. You'll do well to finish by next June."

Pat: "I figured I could work weekends and evenings and still be done by August."

Terry: "I'm sorry, but to be very blunt, you're dreaming. With the demands of a new job and moving this would set you back a lot more than a couple months."

Pat: "Okay, but even if I'm delayed 3 or 4 months don't you think it would be worth it?"

Terry: "Maybe, but what if it turns into a year or two? What if it means you never finish? Believe it or not that happens frequently."

Pat: "So what do you think I should do?"

Terry: "First of all, remember that you have funding here through the end of June so you are not going to be starving. I would start looking for jobs after you have first drafts of all three chapters done; that should be sometime in March. After that you will be in revising mode and doing that while holding down a job should be relatively easy."

Pat: "What if the perfect job comes along and I don't even know that it's available?"

Terry: "At this stage it does make some sense to spend a few minutes each week browsing the job market to get a sense of what's going on. I think you'll see that some kind of job with the DOC opens up pretty regularly."

Pat: "Okay. I have to admit that last night was the first time I've looked at job listings since I started here, so that DOC job was just the first to catch my eye."

Naturally there are exceptions to the generalizations we are painting here; for example, in rare cases a job might be held open for you for many months while you finish your degree. Furthermore, we do not have statistics on the risk involved in leaving for a job too soon and never finishing compared with the risk of finishing and being without a long-term job for some time, but our experiences and the anecdotes of our colleagues suggest that the former is a much larger problem than the latter.

Among my students who took jobs before finishing their degrees only one of them still finished even close to the original schedule. One guy finished 3 years later and only after taking a 9-month leave of absence from his job. Two of them never finished. The worst case was a woman who had her thesis all written except for a Discussion section that she could have produced in less than a week. She left for a job and it was 4 years before she sat down to write that Discussion section. By that time, with the changes in the literature and her drifting away from the subject area, the task had mushroomed from a few days to a few weeks and she gave up. Today she is still working in conservation but at a series of poorly paid, temporary jobs; she would probably be earning twice the salary in a permanent post if she had finished her degree. On the other side of the coin, all of my nearly 50 graduate students have found permanent, professional employment in conservation. The most difficult

*job search took one student almost a year and a half because she elim-
inated two-thirds of the country from her job search, but during that
time she had a series of short-term conservation jobs that were good
experiences for her and she eventually landed a great job close to home.*

Short-term jobs can often create a nice buffer between finishing your degree
and taking up a long-term job. Ideally these will be at or near the university
to facilitate working on publications with your advisor and keeping your local
network vibrant. In fact, many universities employ large numbers of recent
graduates who are in transition to long-term jobs.

In summary, keep your finger on the pulse of the job market, but unless a truly
perfect job comes along, do not start applying for positions until you are close
to finishing your degree. The prospect of having to take a short-term job or two
while landing a long-term job is less risky than having a long-term job severely
delay or prevent finishing your degree, and thereby potentially compromising
your life-long career development.

HOW TO APPLY

You probably already have some aptitude for applications, especially if you won a place at graduate school, and fortunately the key elements of a strong job application are largely the same as for a graduate application, as described in Chapter 4. Helping employers and employees find each other is a major enterprise and many websites can provide assistance in this process.[9.2] Most universities also have career counseling centers that can be very useful, but the best advice will probably come from your network of people who know the culture of conservation careers. Here we provide our thoughts on some major issues.

Written material

Almost all applications will require two fundamental items: a letter of application and a summary of your qualifications, especially your education and work experience. These summaries are called either a résumé or a curriculum vitae (CV for short); a CV is generally longer and more detailed than a résumé (e.g., listing all the conferences you have attended and the title of each presentation you made).

A superior letter of application will explain why you want this job and why your skills make you a strong candidate in a style that exudes confidence, competence, and your unique personality, all without bragging.

For your letter, there are two key issues you need to address: why you want this job and why your skills make you a strong candidate. The art of writing a superior letter of application is to answer these questions without simply repackaging your résumé and to do so in a style that exudes confidence and competence without bragging. Most job announcements come with a detailed list of duties and qualifications and you should make sure that these are addressed in some fashion (see Box 9.1). In some cases this can be succinct and implicit; for example, if it mentions you must have a driver's license you can have a line in your résumé about your experience with various types of vehicles.

Box 9.1 *Job announcements*

Here are a couple of fictitious job announcements to give you some idea of their key elements. These are shorter than most real notices.

Job title: Field Assistant (full time, temporary: June to August)
Organization: Conservation Services, Inc. is a consulting company with 25 permanent staff based in Capital City that undertakes natural resource inventories throughout the nation for a variety of private and public clients.
Responsibilities: Conduct surveys of farmers in the Western River Valley to quantify their attitudes towards proposals for dam removals.
Qualifications: Ability to drive alone in rural areas, communicate with people, and collect data accurately and systematically required. Experience with social surveys and computerized data entry and management desirable.
Compensation: $1800 per month. Housing and a vehicle will be provided.
Application: Send résumé, cover letter, and three reference letters; deadline January 15.

Job title: Policy Director (full time, permanent)
Organization: The Environmental Action Coalition (EAC) is a membership-supported non-profit organization that is dedicated to improving the nation's environmental quality, chiefly by bringing scientific input into the policy-making arena. We have 20,000 members, a staff of 15 based in Capital City and three regional offices, and an annual budget of US$2.7 million.
Responsibilities: Oversee all aspects of EAC's policy initiatives. This will require assessing current environmental issues and the opinions of our members and stakeholders, coordinating activities with diverse collaborators, developing new initiatives, seeking financial support, managing a budget, supervising three staff members, and supporting the Executive Director and EAC activities in general.
Qualifications: We seek someone with a vision and the drive to realize that vision. Experience with environmental policy issues, and strong communication and interpersonal skills are required. Administrative, leadership, and fund-raising skills are highly desirable. A university degree in a relevant discipline is required; a graduate degree is desirable.
Compensation: A competitive salary and benefits package will be offered.
Application: Send a curriculum vitae, a letter that outlines your vision for the future of national environmental policy, and contact information for at least three references. Review of applications will begin March 1.

You may find some qualifications listed that you do not meet, but these should not necessarily dissuade you from applying. Job descriptions are often a portrait of the ideal candidate, which no one will match perfectly. You may decide to ignore the issue and let it arise in the interview if you get that far, but it may be preferable to address the issue head on: "Although familiarity with Pacific algae is listed as a requirement, I am confident that my extensive work with Atlantic algae will allow me to learn the Pacific flora quickly." Conveying all this eloquently and concisely is no easy task, so leave time for multiple drafts with review by critical readers.

Assembling a résumé is much easier than a letter of application: a concise, accurate, professional-looking (no spelling errors!) compilation of your activities will generally suffice. It needs to strongly emphasize professional activities (do not bother mentioning that you played trumpet in your secondary school band) and education (possibly listing the courses most relevant to a particular job), with perhaps a brief mention of relevant recreational activities that indicate you are comfortable in the field (such as scuba diving or mountaineering) and physically fit. Because they are considered confidential you are not likely to be able to peruse a collection of résumés in your department office, but you could ask friends and colleagues to see theirs and you can find samples on the Internet.[9.2]

You may also be asked to provide course transcripts and writing samples; you probably will not need to submit standardized test scores. For government agencies, you are likely to have to fill out elaborate forms required for compliance with stringent government recruitment procedures. You may be inclined to show how witty you are when asked questions like – "Have you ever sworn an oath of allegiance to another sovereign government?" – but resist the temptation. These questions are serious, and your application may first be filtered by personnel who are not considering your professional qualifications.

One fundamental issue is following the specified protocol very carefully and scanning ruthlessly for errors of spelling or grammar. If the applicant pool is large it is easy to reject a poorly prepared application, or to assume that the applicant was not sufficiently interested in the job to invest adequate preparation time. Proper preparation also means uniquely crafting each application, not just changing a few sentences in the application letter. Obviously the application letter really needs to be sculpted to fit the job and even your résumé may need some fine-tuning, for example, by changing the order of items presented to emphasize the skills most relevant to a particular position.

References

Your advisor and members of your advisory committee are obvious candidates to provide references. Additionally, it is often good to include someone from outside of academia; for example, someone who will be using your results to implement conservation action. The exact mix of references you solicit might depend on the nature of the job. Imagine that you were applying for a job with a government agency in the neighboring province; it would be very useful to have a letter from someone in your province's analogous agency saying that you are just the kind of person that is needed for this type of work. You may want to investigate whether anyone in your social network knows someone of influence in the organizations where you are seeking employment. Strive to cultivate good references over an extended period, as mentioned in Chapters 2 and 4. You want people who can write enthusiastically about multiple aspects of your life, not just your performance in a course and what a pleasant colleague you are. It would be useful to ask potential references, especially your advisor, how they perceive your strengths and weaknesses, both as an exercise in identifying potential issues and to assess the impact of a reference letter from this person.

Your references may need to provide letters at the time of application, write letters after you have become a finalist, submit evaluations through online application portals, or be willing to answer questions about you via phone calls or e-mails. In all cases it is a good idea to provide them with three items: your application letter, your résumé, and a copy of the job announcement or hyperlink to it. Make sure you give them plenty of notice in advance. The potential for reference fatigue is one reason for targeting your search to some degree. A shotgun blast of applications wastes everyone's time, including your own.

Interviews

Depending on the culture of the organization and the geographic distribution of applicants, interviews may be by phone, in person, or first by phone and then in person for finalists. They may last 30 minutes or 2 days. They may resemble a casual conversation, or they may be highly structured around a set of questions that will be identical for all applicants. They may involve one interviewer or dozens. You may feel like the process is highly individualized or you may feel like "one little fish" in a large applicant pool.

In the face of this diversity, it is difficult to offer very specific advice, but some generalizations are worth making. Two main attributes that come to the fore in most professional job interviews are being knowledgeable and articulate. Of course, many other factors can come into play – congeniality, creativity, and capacity for hard work to name just three – with the relative importance varying widely among jobs.

Being knowledgeable and articulate is distilled from your years of education and experience, but there are things you can do to polish these characteristics in the context of a particular job interview. First, it is important to obtain some basic background information about the job and the organization. Try browsing the Internet and talking to any contacts you or your advisor might have (i.e., people who know the organization well, but are not involved in the hiring process). After you know the lay of the land, you can initiate a concerted effort to learn more about the subject area of the job and the key people, particularly about those who will interview you.

> One of our students got an interview for his dream job: coordinator for waterfowl conservation back in his home state. He knew this was his big opportunity and he decided to go all out with his preparations. He spent most of 2 weeks reading everything he could find about the state's waterfowl species, conservation plans, hunting laws, and wetland management and he went into the interview probably knowing more about the topic than the people doing the interview. He got the job.

It is easier to be articulate when you are knowledgeable, but that is not the whole story. You will be able to express yourself more clearly with some practice, so ask your friends or advisor to do a mock interview with you. You can certainly anticipate some nearly universal questions, such as "Why do you want this job?" or "What are your strongest and weakest attributes with respect to this position?" Another common question is "Where do you envision yourself in 5 or 10 years?" There may be a hidden agenda here because a prospective employer would often wish to know whether you are likely to stay in the position for a meaningful length of time. With your background work on the position you should be able to predict some more specific questions. Thinking carefully about how you would undertake the job's duties is likely to generate some good ideas that you can weave into your answers. Showing your capacity for innovation may be well received, but be careful to approach this with humility so that you do not appear critical of the organization.

You will probably be asked whether you have any questions, so it would be wise to have some thoughtful ones prepared – not "How many weeks of annual leave would I receive?" Indeed, you should approach the interview process as a two-way street in which you are learning about your potential employer and colleagues to help you make a decision about accepting an offer. This may be difficult to do adequately if the interview is short, but do the best you can and you can always ask questions after you have received an offer and before you accept it.

It may be important that you are perceived to be someone who will fit into the organization. This is an issue with many dimensions – personality, recreational interests, political leanings, wardrobe, hairstyle, and more – and you should try to discuss it with any contacts you have. Clearly you are not a chameleon who will change your fundamental persona just to win a job (e.g., showing up at an interview wearing a campaign button for the conservatives if you are a liberal).

Before an interview learn about the culture of the organization and strive to fit in as best you can, short of becoming a chameleon.

On the other hand, you would not want to lose a great job because your t-shirt and jeans were judged inappropriate. It is sensible to find out what the typical attire for the organization is, and to hedge your bets by dressing one notch more formally.

After the interview, it is a good idea to follow up with a brief e-mail or phone call expressing gratitude for the opportunity to interview, and (assuming you still want the job) reiterating your positive impressions of the organization and indicating that you are still excited about the opportunity.

ACCEPTING A JOB

Some days or weeks after an interview you will probably receive a phone call offering you the post, or explaining that it was offered to someone else. Naturally, if you are not offered the position you need to be polite; you may be back again interviewing with this same person later. It is sensible to ask for feedback in an inquisitive, not accusatory manner: "Can you give me any insights about how my application could have been stronger?"

If you have developed strong skills and have a modicum of flexibility, you will find a rewarding job in conservation.

There is a good chance that you will not be selected after your first couple of interviews, so it is important to capitalize on them as learning experiences.

If you are offered the post, you may be in a position to accept immediately, especially if you have a good idea of the job parameters such as salary and benefits and have already concluded that this is your best option. Commonly things are more complicated and you may want to negotiate certain aspects of the job, or seek more information with which to weigh this job against other options. (Recall from Chapter 4, the delicate issue of deciding between alternative opportunities when you have one bird in the hand, but a more desirable bird still in the bush.)

After 16 years working for various environmental groups I tell newcomers: don't expect heaven; don't accept hell. No job is perfect, and this seems to be especially true for entry-level positions. You shouldn't expect your first job to be a constant adrenalin rush of intellectual stimulation and power politics. But, on the whole, it should be fairly interesting and challenging, and if it's not, then move on. Don't accept a position you know you will hate with the hope that it will evolve into a more interesting one.

Pat: "Thanks, I really am pleased to be offered this job. It's exactly the kind of work I want to do and you have a great organization … "

Kim Green:	"I feel a 'but' coming."
Pat:	"Well, the problem is that my partner Chris is also looking for a job now and we are worried about how that will go in a small town like yours."
Kim Green:	"After you mentioned that you are married to a hydrologist during lunch, I took the liberty of talking to someone in our water resources division and they may have two new positions later this month. We can't guarantee anything of course but if you could send along Chris' résumé they could give us some feedback on what they think the odds are."
Pat:	"Great; I can get Chris' résumé to you this afternoon."
Kim Green:	"Okay, we will get you some feedback from the water folks by Friday. I will need your decision by early next week. Are there any other issues we need to discuss?"
Pat:	"I do wish I could negotiate for a bit more annual leave, but I guess that is fixed by policy high up in the organization."
Kim Green:	"That's right; but if you need more time for your canoe trips it's not hard to get an extra week of leave without pay. We're flexible as long as you get the work done."

And so it goes. Life is always a balancing act in which you have to weigh various options. As long as all parties are open and straightforward, things have a way of working out. Again, if you have developed strong skills and have a modicum of flexibility, you will find a rewarding job in conservation.

Further readings and notes

9.1 The generic job websites such as jobs.com list significant numbers of conservation jobs, but it is probably best to start with ones that are more specialized, such as www.earthworks-jobs.com, www.ecoemploy.com, www.ecojobs.com, eco.org, or www.envirojobs.com.au. Even more specific (and likely to be attuned to your special interests) are the job websites that most professional societies maintain. For example, if you are mainly interested in conservation biology jobs, then check the Society for Conservation Biology site: www.conbio.org/professional-development/scb-job-board. See note 7.2 on finding professional society websites. There also many useful sites that you may find only by word of mouth, so ask for guidance or try searching online for "jobs" or "employment" and one or two key words that cover your interests. At a generic site you may need to try a variety of key words to design a net with a mesh that is neither too coarse nor too fine. The geographic coverage of websites is uneven. Some only cover one country; many that purport to be global in scope are heavily biased toward one country or continent.

9.2 Many job-search websites contain resources that could be useful for your search, such as sample cover letters and résumés. Here are three to check out: www.cvtips .com (available in several languages), www.quintcareers.com/career_resources .html, and sciencecareers.sciencemag.org/tools_tips/how_to_series/how_to_craft_ a_winning_resume.

10 Making a difference

You may have been inspired to read this book because you fancied yourself joining the league of conservation professionals dedicated to the noble goal of "saving the Earth", in some ways akin to medical doctors for biodiversity. Having successfully completed your degree, how do you actually do this? Does getting a job with the word "conservation" in the title and showing up to work every day on time count? Does that make a difference? Well, saving the Earth is not a simple task and, honestly, it requires the dedication of people whose job is more than just a job, people who cultivate their passion for the natural world and integrate their skills and values into the fabric of society.

You can contribute to conservation through multiple avenues – some that you may have traveled as a student or in a conservation apprenticeship, and others that will become available to you after you graduate. Here, we give some suggestions for making a difference during your student years and after you have become a conservation professional. But first, a health warning to all potential conservation professionals: beware of contracting Savior Syndrome.

SAVIOR SYNDROME

Saving the Earth is a lofty goal. You have spent years earning a degree to prepare yourself for your mission, and in some cases donated blood, sweat, and tears to the endeavor. If you have focused on finishing your degree and not had a lot of interaction with the public or conservation stakeholders, you may be susceptible to Savior Syndrome, a disease that often afflicts newly-minted

Saving the Earth as a Career: Advice on Becoming a Conservation Professional, Second Edition. Malcolm L. Hunter, Jr., David B. Lindenmayer, and Aram J. K. Calhoun.
© 2016 John Wiley & Sons, Ltd. Published 2016 by John Wiley & Sons, Ltd.

professionals and may even strike mid-career conservation professionals with well-endowed egos. Symptoms vary in severity, but generally include:

- a belief that the public will bow to your academic credentials and, after showing due respect, will chisel your recommendations into stone tablets;
- a belief that conservation problems can be solved from the top down (the top being people with advanced degrees);
- a belief that having the best-available science, presented by the best scientists (such as yourself), is the key to effective conservation;
- a belief that solutions to conservation issues will only involve experts from a single discipline or that your "way of knowing" is superior to others;
- a belief that that traditional knowledge or practical experience is inferior to academic knowledge;
- a belief that your degree or training precludes you from "getting your boots dirty" on occasion
- a belief that science theory is easily translated into practice;
- a belief that everyone will (or at least should) embrace your value system.

If you have one or more of these symptoms, you may be in the early stages of this syndrome. If untreated, the result is that you will *not* make a difference and will probably find yourself working alone in the face of problems that require collaboration and partners from many different occupations and disciplines. Luckily, treatment is simple.

- Put your training in perspective. The most important things that you have learned in your degree are the abilities to conduct an independent project, solve conservation problems, think critically about new science or perspectives that you encounter, and hopefully to foster collaborations; you have not received divine knowledge.

- Embrace a multidisciplinary approach to conservation. Effective implementation of conservation strategies usually requires skills from many disciplines, including communication, education, science, law, anthropology, government, resource economics, politics, history, and sociology. In particular, conservation requires diverse knowledge, including local and indigenous knowledge that complements scientific knowledge. It also requires good public-relations skills and diplomacy, as described later in this chapter.

- Engage with all stakeholders. Conservation is most effective when the people it will directly affect are involved in crafting a solution (and in some cases, involved in helping to frame the problem and research questions). Local knowledge is invaluable. You must listen to concerns, address perceived problems, and adjust your course based on the input of diverse people. Often, the stakeholders we are most likely to disagree with are the ones that are most important to have at the table. They may be local citizens or specific interest groups (e.g., farming, forestry, or fishing communities) depending on the nature of the conservation project. A bottom-up approach, or at least a meet-in-the-middle hybrid approach, although a slower process than top-down dictatorship, leads to a working compromise or consensus and hence more robust, long-lived conservation practices. Taking the time to learn from stakeholders often leads to more creative and enduring solutions. This is especially critical when working with communities in cultures different than your own; sometimes these cultures do not have to be thousands of miles apart in order to exist.

- Remember that great leaders are not necessarily people who get everybody else to do what they want. A great leader may be someone who can identify and engage key stakeholders and help people with diverse interests to work together respectfully to solve a difficult problem (i.e., a convenor or a boundary agent). Being an effective leader means bottling up your tendency to dominate, and inviting everyone to honestly confront the issue. Being a leader may also mean that you do as much listening and contemplating as delegating or dictating. And, you may still need to get your boots dirty, regardless of your degree or leadership role.

For years I tried to get people in agencies to listen to my ideas. I really believed things would be better if they "followed me." It took a long time to realize that the bureaucrats were not stupid, that they had their own agency mission, and they were not looking for an outsider to tell them what to do. Gradually I got off my soapbox and sat at the table and asked "How can we work together to solve this problem?" Ironically, I became an effective leader only when I sincerely invited others to lead.

• Match your leadership approach to the context of the conservation issue. For example, you may move from the convenor role (described above) to one of advocacy. An advocate may compromise, but their focus is on pushing hard and stating informed opinions to realize goals. In other words, flexibility on means, not goals, may be an effective strategy. A conservation professional may have to be a convenor or an advocate or both; you need to decide which leadership strategy will be most effective in a given situation. This is another situation where keen knowledge of the human dimensions of a problem can be pivotal in realizing goals. If you focus only on the natural phenomena of interest (and ignore the human interactions or politics), you may miss the forest for the trees.

• Dust off your sense of humor and do not take yourself too seriously. You are bound to meet with resistance by some, and you must learn to be level-headed and good-natured. This approach also helps to defuse awkward or contentious situations, and your humility may even break the ice, encouraging others to voice opinions or express knowledge.

Two conservation professionals were working with a small community in the rural USA on a controversial wetland conservation project. Originally, the two were to share responsibilities for giving public presentations and fielding questions. When it became apparent that many in the town were highly suspicious of the project and very concerned about private property rights, the two adjusted their strategy. The expert with a British accent was taken out of the limelight to allow the expert who had lived in the town to launch the project. The added tension that would have been created by an outsider leading the discussion was eliminated and the townspeople were more willing to listen to the woman who had been an active community member before the project. In this case, degrees were irrelevant and local knowledge and "buy in" were crucial. The key was gaining initial trust. After taking the time to address concerns, both professionals were able to play important roles in achieving the project's goals.

If you have no symptoms of the Savior Syndrome, or if you have successfully completed treatment, you are ready to read on and make a difference.

COMPASSION FATIGUE: THE FLIP SIDE OF THE SAVIOR SYNDROME

Recognizing that you are not going to save the world on your own does not mean, however, that you must compromise your values or become deflated by the scale or enormity of environmental problems. As a conservation professional, you may be the only one at the table who speaks for those who cannot speak for themselves: plants, non-human animals, and ecosystems.

To be an effective conservation professional you have to work with people whose values may be different from yours.

Sometimes this comes at a personal cost and may lead to another syndrome: Compassion Fatigue. This term was coined by Charles Figley and refers to a "disorder that affects those who do their work well."[10.1] It is characterized by deep emotional and physical exhaustion, symptoms resembling depression and post-traumatic stress disorder, and by a shift in a person's optimism about the future and the value of their work. In other words, Compassion Fatigue is often a result of doing your job well and becoming emotionally attached to the endeavor; it is not a sign of weakness. As Rachel Remen says: "You can't walk through water without getting wet." Because most conservationists are passionate about their work, it is important to recognize that their work may take both an emotional and physical toll. Compassion Fatigue can be addressed through making sure that you put energy into taking care of yourself (there is no one strategy, every individual will have to explore what relieves stress themselves, or with a professional). After all, if you become traumatized by your work, you will not be as effective or find joy in your life.

MAKING A DIFFERENCE AS A STUDENT

You can make a difference as a graduate student simply by doing fine research that informs conservation and by publishing the efforts of your work. Undergraduate students can make a difference through activities described below, by seeking summer employment with conservation organizations, or by considering undergraduate research projects in conservation. If you are the type of person who needs to stay focused on one task, you should not feel guilty or fear you lack dedication to the greater issues. You may even have an advisor who strongly advises you against participating in any activities unrelated to your work. If this is the case, don't create more stress by worrying about being a super-student. However, if you are one of those people who has a surfeit of energy and a talent for multitasking, there are many ways you can make a difference while honing your skills as a conservation professional.

Project

Your project has the potential to make a difference, particularly if it directly addresses conservation problems, as was discussed in Chapter 5. Including a practitioner on your advisory committee is one important step in formalizing the link between your work and conservation outcomes. A key issue here is that it will make more of a difference if you publish it. The "gold standard" may be peer-reviewed professional journal, but there are certainly worthwhile alternatives to that as described in Chapter 8. For example, you may be able to write a technical report or a manual that will be directly employed by conservation. Although the pressure of getting a job often takes precedent as you near the end of your degree program (see Chapter 9), it is incumbent upon you to share your findings with the conservation community. Even if your work is not ground-breaking research to be published in *Nature* or *Science*, it is a contribution and is part of the process of improving conservation science and practice.

Even a project with limited relevance to a conservation problem is valuable if it develops the skills you need to become a top-notch professional with a paying job. The ability to analyze and appreciate interactions among factors is important in many disciplines and is readily transferable to conservation practice. Often a student cannot control their study topic, so don't blame yourself if your project is "too basic" as long as it helps you understand how the world works and prepares you for securing meaningful work.

Community outreach

Many opportunities for interacting with people outside of academia may be pursued.[10.2] It is often productive to share research or project ideas with stakeholders or local experts before you solidify your plans. The best input is often found in people not associated with academia. For example, if your project is related to conservation issues in forestry, you should consider offering a presentation on your work to people in the forestry community. If you value their experience and knowledge, you will likely benefit from their input, even at the problem formulation stage. You most certainly will be forced to be clear on how your work affects the local community, and any other issues they perceive. You may be exposed to some practical challenges to your direction that you could not possibly appreciate in an academic approach, and you may reap unexpected benefits including modification of your plan or help with project logistics. Think of this as an early investment in how to craft your "management recommendations" at the conclusion of your project.

> *I had a graduate student working on controlling aquatic invasive species in lakes who engaged lake associations and regulators at every stage of her project through meetings and presentations on her work. She received valuable feedback on her project by hearing what the community saw as major concerns, and by listening to their needs she garnered public support for her project and media recognition through a public television broadcast. Specifically, her presentations to community groups and state environmental organizations led to unexpected financial support for the project and volunteer field labor. It also ensured that her work would be used by the groups most closely working with invasive-species issues.*

Joining a local community group with ties to your area of study can be productive. It may allow you to learn from the group, identify how your project fits with local opportunities, and to contribute what you can.

> *One of my most productive graduate school experiences was attending monthly regional stormwater group meetings. I not only learned important lessons about municipal government and the role of research in municipal government, but I was also able to conduct research for the group and other affiliated organizations.*

You can make a difference by conveying your findings to practitioners. In some circles, it is considered unethical not to do so. Most people do not read articles in professional journals and will not benefit from your work unless you find other outlets to disseminate your findings. You can start by inviting practitioners to your defense seminar or giving tailored seminars to key interest groups. Invite the press to learn about your project; such publicity can lead to additional funding opportunities and volunteer labor. Consider publishing a summary of your work in newsletters or web-based publications of interest groups or producing fact sheets for practitioners that summarize your findings. Keep your audience in mind when writing or presenting; they probably have different reading tastes than your advisory committee.

You can make a difference by conveying your research findings to practitioners.

Community outreach does not need to be tied to your project work. Sharing your expertise, or even a passion for natural history, with school groups or nature clubs, can make a difference by spreading appreciation of the natural world, for example by leading field trips. Many not-for-profit conservation organizations need volunteers. You may find that taking on side projects of direct interest to conservation organizations and agencies has diverse benefits. Building a relationship could lead to further collaborations, new career opportunities, and even funding for the side projects. In addition, these opportunities may help you crystallize your own career goals. Just as important, you may find these activities a stress release and may actually return to your academic work feeling refreshed.

Social media

Contributing your research findings to social media is a wonderful way to reach the public, especially young people (see Box 8.2). Scientists can contribute stories to share with the public about research through Facebook, YouTube, Twitter, and blogs. It is a great public-relations tool and a way to share enthusiasm for the natural world with stakeholders. Whet the interests of potential young conservation professionals with tales from the field or lab, and influence the next vanguard of conservation professionals.

A simple blog-post, complete with engaging photos or GoPro video, gives many messages at once: people and conservation are inextricably intertwined, volunteers are enthusiastic, most projects require help to be successful, and there are creative ways to garner help.

After three years of collecting data on blue-spotted salamanders, this Saturday I celebrated the removal of drift fences from around four local vernal pools. I was aided by nine enthusiastic volunteers from the undergraduate Student Chapter of The Wildlife Society at our university, despite the mud and the promise of hard labor. It felt right to have the students helping with the tear down, as the fences have always been both about salamanders and people. I've spent a lot of time alone or paired with a student at these fences counting amphibians, but I could never have done this work without the contribution of other people.

Campus activities

Involvement in student chapters of professional organizations (Society for Conservation Biology, Agriculture and Applied Economics Association, and many more), or campus activist groups that are trying to undertake local conservation activities, is a good way to balance your academic pursuits with other facets of conservation. You will hone interpersonal, organizational, and leadership skills and perhaps have an early exposure to the challenges associated with applying conservation science to practical pursuits.

Mentoring

The importance of being a role model may seem to be a cliché, but most things are clichés because they are truisms. Undergraduate and secondary school students look up to graduate students, and opportunities for inspiring these potential conservation professionals are enormous – inside and outside of the classroom. Similarly, more advanced undergraduates can mentor incoming students. Opportunities for these mentoring relationships (formal or informal) can be found by joining student organizations or chapters of larger professional organizations. Perhaps most importantly, you should feel responsible for mentoring any assistants who work with you, as described in Chapter 6. You can do this by involving them in discussions of how and why data are to be collected, asking their advice on improving methods, and being sure that you ask them what their interests are so you can provide a balanced and rewarding work experience for them. You want students to appreciate that conservation work can be tedious and uninspiring, but that the results and application of the results can be exciting. It is a disservice to the profession to "burn out" assistants to the point that they switch to a profession in accounting. Some graduate students serve on undergraduate honor's thesis committees. This is a wonderful way to share your research and writing experience with college students.

If you are interested in international conservation, you might consider serving as a mentor for students in countries where English is not the first language. Some professional organizations (such as the Society for Conservation Biology) facilitate student–student mentorship where mentors offer to review abstracts and journal papers, and help students gain access to scientific papers.

MAKING A DIFFERENCE AS A CONSERVATION PROFESSIONAL

Job

We have defined the conservation profession rather broadly (Chapter 1), and thus the nature of your contribution to conservation will vary greatly depending on your job. In the best-case scenario, your job has a clear conservation agenda. You may work for a government agency on endangered species, or as an environmental lawyer employed by a conservation NGO developing cap-and-trade systems for carbon emissions. Even if you have a job that has less of a direct conservation focus, you can still make a difference by the way you approach your work. For example, as a teacher, you can infuse conservation into many aspects of your curriculum. If you conduct environmental assessments for a consulting firm, you can be sure that you do a thorough, scientifically defensible job and never sacrifice your integrity simply to satisfy a client's desire for a specific outcome. Share your passion for the natural environment with your co-workers. Your attitude is contagious and can improve the integrity of your workplace.

In any job, if you adhere to a code of conservation ethics and share your values through example, you will make a difference.

You may find yourself in a job where your conservation values are not perfectly aligned with those of your employer, particularly if leadership posts are heavily influenced by the current set of elected officials for local, state or federal government. If you work for a government agency where politics is never too far away, you may be able to gently influence policy simply by raising issues or presenting scientific information on policies being developed that may not have been pursued by your colleagues.

Share your talents through *volunteering* your time. Your job may put you in a good position to mentor other conservation professionals, especially students (see the section on mentoring above). At some point, you will be in a position to "give a break" to a young conservation professional or facilitate contacts with key people to advance their careers. Consider ways in which you may create experiences or opportunities for others.

I teach a wetland conservation and ecology class and each year, I invite experts in different conservation professions (governmental agency, environmental non-governmental, private consulting) to speak to the class. I have never had anyone say no in the 15 years I have been asking for professional input. They all say how much they like to interact with young people and help them to make career choices by explaining the pros and cons of their work. Each year, one alumnus of the university who co-founded an environmental consulting firm spends a day with

> *my class. Although much of their work is for commercial enterprises, they also do unpaid work for conservation projects, provide internships to newly graduated students, scholarships to undergraduates, and are dedicated to sharing their professional expertise with students and newly emerging conservation professionals. Fifteen students in my class were employed by this firm through the interaction in my class.*

The take-home message is that in any job, if you adhere to a code of conservation ethics (see below) and share your values through example, you will make a difference.

Community outreach

Although nobody would criticize a medical doctor for speaking out on preventing AIDS or other public health issues, some academics believe that conservation scientists lose objectivity if their contributions are not confined to books and peer-reviewed publications. However, standing on the sidelines is more akin to isolation than objectivity. We believe that whatever post you take, whether it be academic, consulting, or working for a non-governmental organization, conservation professionals should be actively involved in making the latest information accessible to the public. Think of this civic duty as an important leg of your conservation stool.[10.2] Your personality will dictate how you do that most effectively; for example, not everyone is comfortable with public speaking or with leading a screaming mob of school children through the forest.

Share your knowledge in ordinary language to reach audiences that matter.

Some academics argue (often correctly) that scientific findings lose their precision and nuance when they are translated into ordinary speech for reporters or other audiences. However, we encourage you to think of sharing your knowledge in ordinary language not as dumbing down your discipline, but as reaching audiences that matter. If you don't translate your knowledge, someone else might, and they probably won't do so as effectively as you can.

- **For the extrovert.** Put a friendly face on conservation by giving engaging talks to the public on conservation issues, interacting with schools and camps, and working with practitioners solving conservation problems. You may become actively involved with research or planning projects outside of your work responsibilities. You may develop a relationship with agency bureaucrats and your elected representatives. Opportunities for this are endless; being extroverted, you probably will have discovered this yourself.

- **For the "herder" personality or organizer.** You are a good candidate for leading committees (e.g., chair of a planning board or an advisory committee for an environmental organization) or creating new committees to work on specific conservation issues.

- **For the introvert.** Work behind the scenes. Popularizing conservation through social media, magazine and newspaper articles, or popular books is a wonderful way to make a difference.[10.3] Your local newspaper will almost always publish your letter on an issue related to your expertise. You may also be a candidate for working on committees or advisory boards related to conservation. And, if you try to channel your energy through passion for your work, you may find a voice you didn't know you had.

- **For the expert.** If you are in academia or other jobs where you have a high public profile, sooner or later your expertise will be recognized and you will be called upon to serve in some professional capacity outside of your job description. Although these demands can be significant, to make a difference you should participate at some level. Your services may be requested for providing expert testimony at hearings, serving on boards of professional societies or conservation organizations, or serving on *ad hoc* advisory committees to address specific conservation challenges.

The bottom line is that the more you interact with your colleagues and the public, the more influence you will have in the conservation world. Politicians have figured this out. A lot of important decisions are shaped by social interactions, whether it be over drinks with the "right" people, lunch, fishing, or a round of golf. If you are socially isolated, you will have less direct influence (unless of course you are Einstein or Darwin, in which case you may have a greater impact by staying in your office working out first principles – that is your call). Extrovert or introvert, there are ways you can contribute.

It can be lonely in the middle

You may find yourself in an intermediate position as you try to reconcile the needs of the tree-huggers with those of the tree-cutters. You are most likely to find yourself in this role if you are an academic or a government agency worker who is expected to consider the needs of all constituents. If you find yourself mediating between opposing interests, your skill at remaining level-headed and conciliatory will go a long way towards solving conservation challenges. The key is not to be struck by the axe in the process (either physically or emotionally).[10.4]

Pat: "Hi Terry, thought I would give you a call."

Terry: "Hi Pat, thanks for calling; it's always great to hear from my star former students. So ... how's your new sustainable forestry job going up there in Barkton?"

Pat: "Well, the job's actually the reason why I thought I would call you, rather than send an e-mail."

Terry: "Uh-oh, sounds ominous."

Pat: "Yes, it's been a tough couple of months. As they say, this job ain't easy."

Terry: "Why is that? I thought Barkton would be a great place to work, great forests and scenery, nice community, affordable housing ... sounds much better than the office that I'm stuck in."

Pat: "Barkton is nice ... but the forest debate here is very polarized. There is quite a group of people here who just hate forest harvesting and strongly believe is should be stopped. They are constantly butting heads with the pro-forestry people, many of whom depend on logging for their livelihoods."

Terry: "Sounds like all the ingredients for a typical resource conflict."

Pat: "Yes, but my job means that I have to develop management plans that try to balance ecosystem integrity and logging. I get hammered by the loggers because they say that I am caving in to the greenies. But then the other side hammers me as well because

they reckon I give in too easily to the loggers. I have to say that it gets pretty lonely being stuck in the middle."

Terry: "I think I understand – although I know that does not make it any easier for you. Of course both sides are going to lobby really hard to get the best outcome for their cause. Try not to take it too personally, and stick to your guns about what your scientific background tells you is the right way forward. It's also a really valuable thing to do, not just around Barkton, but around the world."

Pat: "OK, thanks for the discussion; it's good to talk to someone about these things. In fact, do you know anyone else who serves in this intermediary management role that I could connect with?"

Terry: "Sure. I'll e-mail you some names. And finally, stick with it; you're doing an important job."

LIFE STYLE

If you live in a developed country, the reality is you are probably consuming more than your fair share of resources. Yet there are still things that you can do to ameliorate the impacts of life in an environment of excesses. You know what they are and they need not be elaborated upon here. Not practicing what you preach can tarnish your credibility in the public eye, and may undermine your professional advice/recommendations. If the conservation professionals can't bear to take their own advice or sacrifice for the greater good, who will?

> One of my colleagues is a well-respected ecologist who was featured in a magazine article on global climate change where he enumerated the anthropogenic causes of global warming and exhorted the public to do their part to reduce carbon dioxide emissions by selecting energy-efficient vehicles instead of gas-guzzling sports utility vehicles. After being interviewed for this article, he got into his Ford Explorer SUV and drove to his home, which was walking distance from campus.

We can all be hypocrites at times, but it is good to be aware of the implications of your actions and to do as much as you can without making yourself miserable.

CONSERVATION ETHICS

A code of ethics identifies key principles that should become standard operating procedure for the community for which they were drafted. Many professional organizations have codes of ethics; the Society for Conservation Biology has identified 15 guidelines for conservation professionals (Box 10.1). At least six of these have been covered in our discussion of how you can help to save the Earth. We suggest reading these, thinking about what they mean in the context of your job and your personal life, and revisiting them again in a year to see if you are still meeting the standard. These guidelines would be good discussion items for students and faculty.

By choosing to become a conservational professional you have already taken the first large step towards acting ethically toward your fellow humans and all the other life forms with which we share the Earth.

For some people, the line between ethical and unethical behavior becomes blurred by the exigencies of everyday living, especially financial responsibilities. Laws, principles, and guidelines are codified because human beings do not always act altruistically. It takes integrity and patience to make positive changes in organizations, institutions, and conservation practices, but with due diligence, learning from our past actions, and seeking foresight, we can all accomplish this together. By choosing to become a conservational professional and working hard to make a difference, you have already taken the first large step towards acting ethically toward your fellow humans and all the other life forms with which we share the Earth.

Box 10.1 Society for Conservation Biology Code of Ethics

The mission of the Society for Conservation Biology, a global community of conservation professionals, is to advance the science and practice of conserving the Earth's biological diversity. To meet this goal, we encourage all conservation scientists and practitioners to adhere to the following.

1 Actively disseminate information to promote understanding of and appreciation for biodiversity and the science of conservation biology.

2 Advocate the use of reliable information, rigorous scientific methodology, and credible inference in management decisions affecting biodiversity.

3 Recognize that uncertainty is inherent in managing ecosystems and species and encourage application of the precautionary principle in management and policy decisions affecting biodiversity.

4 Recognize their responsibility to conservation and scientific honesty, and inform other scientists, the public, and prospective clients or employers of this responsibility.

5 Avoid actions or omissions that may compromise their responsibility to conservation and science.

6 Be willing to volunteer their services for the public good at a level appropriate to their financial abilities.

7 Perform professional services or peer reviews only in their areas of competence, cooperate with other professionals in the best interest of conservation, and refer clients to other professionals with appropriate expertise.

8 Refuse to allow personal interests, compensation, or personal relationships to interfere with their professional judgment or advice.

9 Scrupulously avoid plagiarism; acknowledge the limitations of their research design, data, and interpretation of results; disclose conflicts of interest; honestly discuss their findings; and attempt to correct misrepresentation of their research by others.

10 Claim authorship of a publication or report only when they have contributed substantially to the conception, design, data collection, analysis,

(continued)

Box 10.1 Society for Conservation Biology Code of Ethics (continued)

or interpretation, or have helped draft or revise the article, and approve of the published version.

11 When working professionally, especially outside their region of residence, interact and collaborate with counterparts, present seminars, confer regularly with appropriate officials, share information, involve colleagues and students in professional activities, contribute to local capacity-building, and equitably share the benefits arising from the use of local knowledge, practices, and genetic resources.

12 Treat colleagues and professional contacts respectfully and support fair standards of employment and treatment for those engaged in the practice of conservation biology.

13 Work to ensure that no colleague is unjustly deprived of his or her job, reputation, ability to publish, or scientific freedom as a result of his or her conservation efforts.

14 Protect the rights and welfare of human subjects used in research and obtain the informed consent of those subjects.

15 Adhere to the highest standards for treatment of animals used in research in a way that contributes most positively to sustaining natural populations and ecosystems.

Further readings and notes

10.1 Figley, Charles. 1995. *Compassion Fatigue: Coping With Secondary Traumatic Stress Disorder In Those Who Treat The Traumatized*. Routledge Publishing, New York, N.Y.

Matthieu, F. 2012. *The Compassion Fatigue Workbook: Creative Tools for Transforming Compassion Fatigue and Vicarious Traumatization*. Routledge Press, New York, N.Y.

10.2 Jacobson, S.K. 2009. *Communication Skills for Conservation Professionals*. 2nd edn. Island Press, Washington, D.C.

Jacobson, S.K., M.D. McDuff, and M.C. Monroe. 2006. *Conservation Education and Outreach Techniques*. Oxford University Press, New York, N.Y.

10.3 E.O. Wilson is an example of someone who has contributed enormously to conservation through his popular writings. See his autobiography, *Naturalist*, and *The Diversity of Life* and *The Future of Life*.

If you enjoy reading about the lives of naturalists and conservation professionals, you should also have a look at Eric Dinerstein's autobiography, *Tigerland and Other Unintended Destinations* (2005, Island Press), Berndt Heinrich's *A Year in the Maine Woods* (1994, Perseus Books), and Jane Goodall's *In the Shadow of Man* (2010, Houghton Mifflin Harcourt)

10.4 Reich, S. M., and J.A. Reich. 2006. Cultural competence in interdisciplinary collaborations: A method for respecting diversity in research partnerships. *American Journal of Community Psychology*, **38**, 51–62.

Pielke, R.A. 2007. *The Honest Broker. Making Sense of Science in Policy and Politics.* Cambridge University Press, Cambridge, United Kingdom.

Senecah, S. L. 2004. The trinity of voice: The role of practical theory in planning and evaluating the effectiveness of environmental participatory processes. Pages 13–33 in S. Depoe, J. W. Delicath, and M.A. Elsenbeer (Eds.), *Communication and Public Participation in Environmental Decision Making.* SUNY Press, Albany, New York, N.Y.

Index

Saving the Earth as a Career: Advice on Becoming a Conservation Professional, Second Edition.
Malcolm L. Hunter, Jr., David B. Lindenmayer, and Aram J. K. Calhoun.
© 2016 John Wiley & Sons, Ltd. Published 2016 by John Wiley & Sons, Ltd.